THE

CLIMATE

REPORT

THE
CLIMATE
REPORT

The NATIONAL CLIMATE
ASSESSMENT—
IMPACTS, RISKS, *and*
ADAPTATION
in the
UNITED STATES

U.S. Global Change Research Program

MELVILLE HOUSE
BROOKLYN · LONDON

THE CLIMATE REPORT

First published in 2018 by U.S. Government Publishing Office
This report is in the public domain. Some materials used herein are copyrighted and permission
was granted for their publication in this book. For subsequent uses that include such
copyrighted materials, permission for reproduction must be sought from the copyright holder.
In all cases, credit mut be given for copyright materials. All other materials are free
to use with credit to the Fourth National Climate Assessment, Volume II. Full report available
online at: nca2018.globalchange.gov.

First Melville House Printing: January 2019

Melville House Publishing   Melville House UK
46 John Street   and   16/18 Woodford Road
Brooklyn, NY 11201   London E7 0HA

mhpbooks.com
@melvillehouse

ISBN: 978-1-61219-802-6
ISBN: 978-1-61219-803-3 (eBook)

Printed in Canada
1 3 5 7 9 10 8 6 4 2

A catalog record for this book is available from the Library of Congress

Federal Steering Committee

David Reidmiller, Chair, U.S. Global Change Research Program

Benjamin DeAngelo, Vice Chair, Department of Commerce

Farhan Akhtar, Department of State

Daniel Barrie, Department of Commerce

Virginia Burkett, Department of the Interior

Jennifer Carroll, National Science Foundation

Lia Cattaneo, Department of Transportation (through December 2017)

Pierre Comizzoli, Smithsonian Institution

Daniel Dodgen, Department of Health and Human Services

Noel Gurwick, U.S. Agency for International Development

Pat Jacobberger-Jellison, National Aeronautics and Space Administration

Rawlings Miller, Department of Transportation (May – August 2018)

Kurt Preston, Department of Defense

Margaret Walsh, Department of Agriculture

Tristam West, Department of Energy

Darrell Winner, Environmental Protection Agency

Subcommittee on Global Change Research

Virginia Burkett, Acting Chair, Department of the Interior

Gerald Geernaert, Vice Chair, Department of Energy

John Balbus, Department of Health and Human Services

Bill Breed, U.S. Agency for International Development (through February 2018)

Pierre Comizzoli, Smithsonian Institution

Noel Gurwick, U.S. Agency for International Development (since February 2018)

Wayne Higgins, Department of Commerce

Scott Harper, Department of Defense

William Hohenstein, Department of Agriculture

Jack Kaye, National Aeronautics and Space Administration

Dorothy Koch, Department of Energy

Barbara McCann, Department of Transportation

Andrew Miller, Environmental Protection Agency

James Reilly, Department of the Interior

Trigg Talley, Department of State

Maria Uhle, National Science Foundation

Executive Leadership and White House Liaisons

Michael Kuperberg, U.S. Global Change Research Program

David Reidmiller, U.S. Global Change Research Program

Chloe Kontos, Executive Director, National Science and Technology Council

Kimberly Miller, Office of Management and Budget

Administrative Lead Agency

Department of Commerce / National Oceanic and Atmospheric Administration

Recommended Citation

USGCRP, 2018: *Impacts, Risks, and Adaptation in the United States: Fourth National Climate Assessment, Volume II: Report-in-Brief* [Reidmiller, D.R., C.W. Avery, D.R. Easterling, K.E. Kunkel, K.L.M. Lewis, T.K. Maycock, and B.C. Stewart (eds.)]. U.S. Global Change Research Program, Washington, DC, USA, 186 pp.

In August 2018, temperatures soared across the northwestern United States. The heat, combined with dry conditions, contributed to wildfire activity in several states and Canada. The image to the left is of the Howe Ridge Fire from across Lake McDonald in Montana's Glacier National Park on the night of August 12, roughly 24 hours after it was ignited by lightning. The fire spread rapidly, fueled by record-high temperatures and high winds, leading to evacuations and closures of parts of the park. The satellite image below, acquired on August 15, shows plumes of smoke from wildfires on the northwestern edge of Lake McDonald.

Wildfires impact communities throughout the United States each year. In addition to threatening individual safety and property, wildfire can worsen air quality locally and, in many cases, throughout the surrounding region, with substantial public health impacts including increased incidence of respiratory illness (Ch. 13: Air Quality, KM 2; Ch. 14: Health, KM 1; Ch. 26: Alaska, KM 3). As the climate warms, projected increases in wildfire frequency and area burned are expected to drive up costs associated with health effects, loss of homes and infrastructure, and fire suppression (Ch. 6: Forests, KM 1; Ch. 17: Complex Systems, Box 17.4). Increased wildfire activity is also expected to reduce the opportunity for and enjoyment of outdoor recreation activities, affecting quality of life as well as tourist economies (Ch. 7: Ecosystems, KM3; Ch. 13: Air Quality, KM 2; Ch. 14: Tribal, KM 1; Ch. 19: Southeast, KM3; Ch. 24: Northwest, KM 4).

Human-caused climate change, land use, and forest management influence wildfires in complex ways (Ch. 17: Complex Systems, KM 2). Over the last century, fire exclusion policies have resulted in higher fuel availability in most U.S. forests (CSSR, Ch. 8.3, KF 6). Warmer and drier conditions have contributed to an increase in the incidence of large forest fires in the western United States and Interior Alaska since the early 1980s, a trend that is expected to continue as the climate warms and the fire season lengthens (Ch. 1: Overview, Figure 1.2k; CSSR, Ch. 8.3, KF 6). The expansion of human activity into forests and other wildland areas has also increased over the past few decades. As the footprint of human settlement expands, fire risk exposure to people and property is expected to increase further (Ch. 5: Land Changes, KM 2).

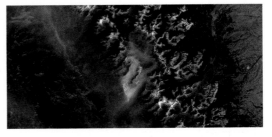

Credits

National Park Service (top); NASA Earth Observatory image by Joshua Stevens, using Landsat data from the U.S. Geological Survey (bottom).

TABLE OF CONTENTS
FOURTH NATIONAL CLIMATE ASSESSMENT REPORT-IN-BRIEF

Front Matter

About this Report 1

Guide to the Report...4

Summary Findings...11

1. Overview 21

What Has Happened Since the Last National Climate Assessment?.... 56

National Topics 63

2. Our Changing Climate 64

3. Water.. 67

4. Energy Supply, Delivery, and Demand 70

5. Land Cover and Land-Use Change 73

6. Forests... 76

7. Ecosystems, Ecosystem Services, and Biodiversity 79

8. Coastal Effects... 82

9. Oceans and Marine Resources 85

10. Agriculture and Rural Communities... 88

11. Built Environment, Urban Systems, and Cities....................... 92

12. Transportation 95

13. Air Quality 98

14. Human Health 101

15. Tribes and Indigenous Peoples 104

16. Climate Effects on U.S. International Interests 107

17. Sector Interactions, Multiple Stressors, and Complex Systems 110

Regions 115

18. Northeast 116

19. Southeast 121

20. U.S. Caribbean 126

21. Midwest 131

22. Northern Great Plains 136

23. Southern Great Plains 141

24. Northwest 144

25. Southwest 148

26. Alaska 153

27. Hawai'i and U.S.-Affiliated Pacific Islands 157

Responses 163

28. Reducing Risks Through Adaptation Actions 164

29. Reducing Risks Through Emissions Mitigation 168

Authors and Contributors 173

Appendix 187

About This Report

The National Climate Assessment

The Global Change Research Act of 1990 mandates that the U.S. Global Change Research Program (USGCRP) deliver a report to Congress and the President no less than every four years that "1) integrates, evaluates, and interprets the findings of the Program . . .; 2) analyzes the effects of global change on the natural environment, agriculture, energy production and use, land and water resources, transportation, human health and welfare, human social systems, and biological diversity; and 3) analyzes current trends in global change, both human-induced and natural, and projects major trends for the subsequent 25 to 100 years."[1]

The Fourth National Climate Assessment (NCA4) fulfills that mandate in two volumes. This report, Volume II, draws on the foundational science described in Volume I, the *Climate Science Special Report* (CSSR).[2] Volume II focuses on the human welfare, societal, and environmental elements of climate change and variability for 10 regions and 18 national topics, with particular attention paid to observed and projected risks, impacts, consideration of risk reduction, and implications under different mitigation pathways. Where possible, NCA4 Volume II provides examples of actions underway in communities across the United States to reduce the risks associated with climate change, increase resilience, and improve livelihoods.

This assessment was written to help inform decision-makers, utility and natural resource managers, public health officials, emergency planners, and other stakeholders by providing a thorough examination of the effects of climate change on the United States.

Climate Science Special Report: NCA4 Volume I

The *Climate Science Special Report* (CSSR), published in 2017, serves as the first volume of NCA4. It provides a detailed analysis of how climate change is affecting the physical earth system across the United States and provides the foundational physical science upon which much of the assessment of impacts in this report is based. The CSSR integrates and evaluates current findings on climate science and discusses the uncertainties associated with these findings. It analyzes trends in climate change, both human-induced and natural, and projects major trends to the end of this century. Projected changes in temperature, precipitation patterns, sea level rise, and other climate outcomes are based on a range of scenarios widely used in the climate research community, referred to as Representative Concentration Pathways (RCPs). As an assessment and analysis of the physical science, the CSSR provides important input to the development of other parts of NCA4 and their primary focus on the human welfare, societal, economic, and environmental elements of climate change. A summary of the CSSR is provided in Chapter 2 (Our Changing Climate) of this report; the full report can be accessed at science2017.globalchange.gov.

About the Report-in-Brief

The NCA4 Volume II Report-in-Brief presents overall Summary Findings, an Overview that synthesizes material from the underlying chapters, and Executive Summaries for each chapter of this volume.

The 186-page Report-in-Brief is available as a downloadable PDF at https://nca2018.globalchange.gov/downloads.

Report Development, Review, and Approval Process

The National Oceanic and Atmospheric Administration (NOAA) served as the administrative lead agency for the preparation of this report. A Federal Steering Committee, composed of representatives from USGCRP agencies, oversaw the report's development.

A team of more than 300 federal and non-federal experts—including individuals from federal, state, and local governments, tribes and Indigenous communities, national laboratories, universities, and the private sector—volunteered their time to produce the assessment, with input from external stakeholders at each stage of the process. A series of regional engagement workshops reached more than 1,000 individuals in over 40 cities, while listening sessions, webinars, and public comment periods provided valuable input to the authors. Participants included decision-makers from the public and private sectors, resource and environmental managers, scientists, educators, representatives from businesses and nongovernmental organizations, and the interested public.

NCA4 Volume II was thoroughly reviewed by external experts and the general public, as well as the Federal Government (that is, the NCA4 Federal Steering Committee and several rounds of technical and policy review by the 13 federal agencies of the USGCRP). An expert external peer review of the whole report was performed by an ad hoc committee of the National Academies of Sciences, Engineering, and Medicine (NASEM).[3] Additional information on the development of this assessment can be found in Appendix 1: Report Development Process.

Sources Used in This Report

The findings in this report are based on an assessment of the peer-reviewed scientific literature, complemented by other sources (such as gray literature) where appropriate. In addition, authors used well-established and carefully evaluated observational and modeling datasets, technical input reports, USGCRP's sustained assessment products, and a suite of scenario products. Each source was determined to meet the standards of the Information Quality Act (see Appendix 2: Information in the Fourth National Climate Assessment).

Sustained Assessment Products

The USGCRP's sustained assessment process facilitates and draws upon the ongoing participation of scientists and stakeholders, enabling the assessment of new information and insights as they emerge. The USGCRP led the development of two major sustained assessment products as inputs to NCA4: *The Impacts of Climate Change on Human Health in the United States: A Scientific Assessment*[4] and the *Second State of the Carbon Cycle Report*.[5] In addition, USGCRP agencies contributed products that improve the thoroughness of this assessment, including the U.S. Department of Agriculture's scientific assessment *Climate Change, Global Food Security, and the U.S. Food System*;[6] NOAA's Climate Resilience Tool Kit, Climate Explorer, and State Climate Summaries; the U.S. Environmental Protection Agency's updated economic impacts of climate change report;[7] and a variety of USGCRP indicators and scenario products that support the evaluation of climate-related risks (see Appendix 3: Data Tools and Scenario Products).

USGCRP Scenario Products

As part of the sustained assessment process, federal interagency groups developed a suite of high-resolution scenario products that span a range of plausible future changes (through at least 2100) in key environmental parameters. This new generation of USGCRP scenario products (hosted at https://scenarios.globalchange. gov) includes

- changes in average and extreme statistics of key climate variables (for example, temperature and precipitation),

- changes in local sea level rise along the entire U.S. coastline,

- changes in population as a function of demographic shifts and migration, and

- changes in land use driven by population changes.

USGCRP scenario products help ensure consistency in underlying assumptions across the report and therefore improve the ability to compare and synthesize results across chapters. Where possible, authors have used the range of these scenario products to frame uncertainty in future climate and associated effects as it relates to the risks that are the focus of their chapters. As discussed briefly elsewhere in this Front Matter and in more detail in Appendix 3 (Data Tools and Scenario Products), future scenarios referred to as RCPs provide the global framing for NCA4 Volumes I and II. RCPs focus on outputs (such as emissions and concentrations of greenhouse gases and particulate matter) that are in turn fed into climate models. As such, a wide range of future socioeconomic assumptions, at the global and national scale (such as population growth, technological innovation, and carbon intensity of energy mix), could be consistent with the RCPs used throughout NCA4. For this reason, further guidance on U.S. population and land-use assumptions was provided to authors. See Appendix 3: Data Tools and Scenario Products, including Table A3.1, for additional detail on these scenario products.

Guide to the Report

Summary Findings

The 12 Summary Findings represent a very high-level synthesis of the material in the underlying report. They consolidate Key Messages and supporting evidence from 16 underlying national-level topic chapters, 10 regional chapters, and 2 response chapters.

Overview

The Overview presents the major findings alongside selected highlights from NCA4 Volume II, providing a synthesis of material from the underlying report chapters.

Chapter Text

Key Messages and Traceable Accounts

Chapters are centered around Key Messages, which are based on the authors' expert judgment of the synthesis of the assessed literature. With a view to presenting technical information in a manner more accessible to a broad audience, this report aims to present findings in the context of risks to natural and/or human systems. Assessing the risks to the Nation posed by climate change and the measures that can be taken to minimize those risks helps users weigh the consequences of complex decisions.

Since risk can most meaningfully be defined in relation to objectives or societal values, Key Messages in each chapter of this report aim to provide answers to specific questions about what is at risk in a particular region or sector and in what way. The text supporting each Key Message provides evidence, discusses implications, identifies intersections between systems or cascading hazards, and points out paths to greater resilience. Where a Key Message focuses on managing risk, authors considered the following questions:

- What do we value? What is at risk?

- What outcomes do we wish to avoid with respect to these valued things?

- What do we expect to happen in the absence of adaptive action and/or mitigation?

- How bad could things plausibly get? Are there important thresholds or tipping points in the unique context of a given region, sector, and so on?

These considerations are encapsulated in a single question: What keeps you up at night? Importantly, climate is only one of many drivers of change and risk. Where possible, chapters provide information about the dominant sources of uncertainty (such as scientific uncertainty or socioeconomic factors), as well as information regarding other relevant non-climate stressors.

Each Key Message is accompanied by a Traceable Account that restates the Key Message found in the chapter text with calibrated confidence and likelihood language (see Table 1). These Traceable Accounts also document the supporting evidence and rationale the authors used in reaching their conclusions, while also providing information on sources of uncertainty. More information on Traceable Accounts is provided below.

Our Changing Climate

USGCRP oversaw the production of the *Climate Science Special Report* (CSSR): NCA4 Volume I,[2] which assesses the current state of science relating to climate change and its physical impacts. The CSSR is a detailed analysis of how climate change affects the physical earth system across the United States. It presents foundational information and projections for climate change that improve consistency across analyses in NCA4 Volume II. The CSSR is the basis for the physical climate science summary presented in Chapter 2 (Our Changing Climate) of this report.

National Topic Chapters

The national topic chapters summarize current and future climate change related risks and what can be done to reduce those risks. These national chapters also synthesize relevant content from the regional chapters. New national topic chapters for NCA4 include Chapter 13: Air Quality; Chapter 16: Climate Effects on U.S. International Interests; and Chapter 17: Sector Interactions, Multiple Stressors, and Complex Systems.

Regional Chapters

Responding to public demand for more localized information—and because impacts and adaptation tend to be realized at a more local level—NCA4 provides greater detail in the regional chapters compared to the national topic chapters. The regional chapters assess current and future risks posed by climate change to each of NCA4's 10 regions (see Figure 1) and what can be done to minimize risk. Challenges, opportunities, and success stories for managing risk are illustrated through case studies.

National Climate Assessment Regions

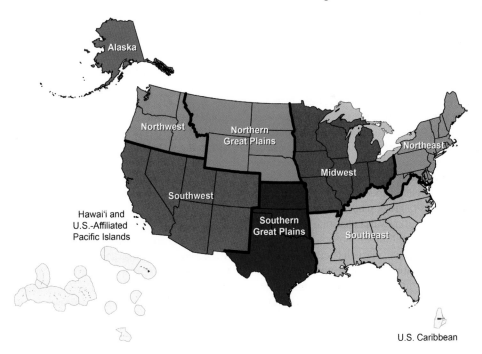

Figure 1: Map of the ten regions used throughout NCA4.

The regions defined in NCA4 are similar to those used in the Third National Climate Assessment (NCA3),[8] with these exceptions: the Great Plains region, formerly stretching from the border of Canada to the border of Mexico, is now divided into the Northern Great Plains and Southern Great Plains along the Nebraska–Kansas border; and content related to the U.S. Caribbean islands is now found in its own chapter, distinct from the Southeast region.

Response Chapters

The response chapters assess the science of adaptation and mitigation, including benefits, tradeoffs, and best practices of ongoing adaptation measures and quantification of economic damages that can be avoided by reducing greenhouse gas emissions. The National Climate Assessment does not evaluate or recommend specific policies.

Economic Estimates

To the extent possible, economic estimates in this report have been converted to 2015 dollars using the U.S. Bureau of Economic Affairs' Implicit Price Deflators for Gross Domestic Product, Table 1.1.9. For more information, please visit: https://bea.gov/national/index.htm. Where documented in the underlying literature, discount rates in specific estimates in this assessment are noted next to those projections.

Use of Scenarios

Climate modeling experts develop climate projections for a range of plausible futures. These projections capture variables such as the relationship between human choices, greenhouse gas (GHG) and particulate matter emissions, GHG concentrations in our atmosphere, and the resulting impacts, including temperature change and sea level rise. Some projections are consistent with continued dependence on fossil fuels, while others are achieved by reducing

GHG emissions. The resulting range of projections reflects, in part, the uncertainty that comes with quantifying future human activities and their influence on climate.

The most recent set of climate projections developed by the international scientific community is classified under four Representative Concentration Pathways, or RCPs.[9] A wide range of future socioeconomic assumptions could be consistent with the RCPs used throughout NCA4.

NCA4 focuses on RCP8.5 as a "higher" scenario, associated with more warming, and RCP4.5 as a "lower" scenario with less warming. Other RCP scenarios (e.g., RCP2.6, a "very low" scenario) are used where instructive, such as in analyses of mitigation science issues. To promote understanding while capturing the context of the RCPs, authors use the phrases "a higher scenario (RCP8.5)" and "a lower scenario (RCP4.5)." RCP8.5 is generally associated with higher population growth, less technological innovation, and higher carbon intensity of the global energy mix. RCP4.5 is generally associated with lower population growth, more technological innovation, and lower carbon intensity of the global energy mix. NCA4 does not evaluate the feasibility of the socioeconomic assumptions within the RCPs. Future socioeconomic conditions—and especially the relationship between economic growth, population growth, and innovation—will have a significant impact on which climate change scenario is realized. The use of RCP8.5 and RCP4.5 as core scenarios is broadly consistent with the range used in NCA3.[8] For additional detail on these scenarios and what they represent, please see Appendix 3 (Data Tools and Scenario Products), as well as Chapter 4 of the *Climate Science Special Report*.[10]

Treatment of Uncertainties: Risk Framing, Confidence, and Likelihood

Risk Framing

In March 2016, NASEM convened a workshop, Characterizing Risk in Climate Change Assessments, to assist NCA4 authors in their analyses of climate-related risks across the United States.[11] To help ensure consistency and readability across chapters, USGCRP developed guidance on communicating the risks and opportunities that climate change presents, including the treatment of scientific uncertainties. Where supported by the underlying literature, authors were encouraged to

- describe the full scope of potential climate change impacts, both negative and positive, including more extreme impacts that are less likely but would have severe consequences, and communicate the range of potential impacts and their probabilities of occurrence;

- describe the likelihood of the consequences associated with the range of potential impacts, the character and quality of the consequences, both negative and positive, and the strength of available evidence;

- communicate cascading effects among and within complex systems; and

- quantify risks that could be avoided by taking action.

Additional detail on how risk is defined for this report, as well as how risk-based framing was used, is available in Chapter 1: Overview (see Box 1.2: Evaluating Risks to Inform Decisions).

Traceable Accounts: Confidence and Likelihood

Throughout NCA4's assessment of climate-related risks and impacts, authors evaluated the range of information in the scientific literature to the fullest extent possible, arriving at a series of Key Messages for each chapter. Drawing on guidance developed by the Intergovernmental Panel on Climate Change (IPCC),[12] chapter authors further described the overall reliability in their conclusions using these metrics in their chapter's Traceable Accounts:

- **Confidence** in the validity of a finding based on the type, amount, quality, strength, and consistency of evidence (such as mechanistic understanding, theory, data, models, and expert judgment); the skill, range, and consistency of model projections; and the degree of agreement within the body of literature.

- **Likelihood**, which is based on measures of uncertainty expressed probabilistically (in other words, based on statistical analysis of observations or model results or on the authors' expert judgment).

The author team's expert assessment of confidence for each Key Message is presented in the chapter's Traceable Accounts. Where the authors consider it is scientifically justified to report the likelihood of a particular impact within the range of possible outcomes, Key Messages in the Traceable Accounts also include a likelihood designation. Traceable Accounts describe the process and rationale the authors used in reaching their conclusions, as well as their confidence in these conclusions. They provide additional information about the quality of information used and allow traceability to data and resources.

Confidence Level
Very High
Strong evidence (established theory, multiple sources, confident results, well-documented and accepted methods, etc.), high consesus
High
Moderate evidence (several sources, some consistency, methods vary and/or documentation limited, etc.), medium consensus
Medium
Suggestive evidence (a few sources, limited consistency, models incomplete, methods emerging, etc.), competing schools of thought
Low
Inconclusive evidence (limited sources, extrapolations, inconsistent findings, poor documentation and/or methods not tested, etc.), disagreement or lack of opinions among experts

Likelihood				
Very Likely	**Likely**	**As Likely as Not**	**Unlikely**	**Very Unlikely**
≥ 9 in 10	≥ 2 in 3	= 1 in 2	≤ 1 in 3	≤ 1 in 10

Table 1: This table describes the meaning of the various categories of confidence level and likelihood assessment used in NCA4. The levels of confidence are the same as they appear in the CSSR (NCA4 Volume I). And while the likelihood scale is consistent with the CSSR, there are fewer categories, as that report relies more heavily on quantitative methods and statistics. This "binning" of likelihood is consistent with other USGCRP sustained assessment products, such as the Climate and Health Assessment[4] and NCA3.[8]

Glossary of Terms

NCA4 uses the glossary available on the USGCRP website (http://www.globalchange.gov/climate-change/glossary). It was developed for NCA3 and largely draws from the IPCC glossary of terms. Over time, it has been updated with selected new terms from more recent USGCRP assessments, including *The Impacts of Climate Change on Human Health in the United States* (https://health2016.globalchange.gov/glossary-and-acronyms) and the *Climate Science Special Report* (https://science2017.globalchange.gov/chapter/appendix-e/).

References

1. Global Change Research Act of 1990. Pub. L. No. 101-606, 104 Stat. 3096-3104, November 16, 1990. http://www.gpo.gov/fdsys/pkg/STATUTE-104/pdf/STATUTE-104-Pg3096.pdf

2. USGCRP, 2017: Climate Science Special Report: Fourth National Climate Assessment, Volume I. Wuebbles, D.J., D.W. Fahey, K.A. Hibbard, D.J. Dokken, B.C. Stewart, and T.K. Maycock, Eds. U.S. Global Change Research Program, Washington, DC, USA, 470 pp. http://dx.doi.org/10.7930/J0J964J6

3. National Academies of Sciences, Engineering, and Medicine, 2018: *Review of the Draft Fourth National Climate Assessment*. The National Academies Press, Washington, DC, 206 pp. http://dx.doi.org/10.17226/25013

4. USGCRP, 2016: *The Impacts of Climate Change on Human Health in the United States: A Scientific Assessment*. U.S. Global Change Research Program, Washington, DC, 312 pp. http://dx.doi.org/10.7930/J0R49NQX

5. USGCRP, 2018: Second State of the Carbon Cycle Report (SOCCR2): A Sustained Assessment Report. Cavallaro, N., G. Shrestha, R. Birdsey, M. Mayes, R. Najjar, S. Reed, P. Romero-Lankao, and Z. Zhu, Eds. U.S. Global Change Research Program, Washington, DC, 877 pp. http://dx.doi.org/10.7930/SOCCR2.2018

6. Brown, M.E., J.M. Antle, P. Backlund, E.R. Carr, W.E. Easterling, M.K. Walsh, C. Ammann, W. Attavanich, C.B. Barrett, M.F. Bellemare, V. Dancheck, C. Funk, K. Grace, J.S.I. Ingram, H. Jiang, H. Maletta, T. Mata, A. Murray, M. Ngugi, D. Ojima, B. O'Neill, and C. Tebaldi, 2015: Climate Change, Global Food Security, and the U.S. Food System. U.S. Global Change Research Program, Washington, DC, 146 pp. http://dx.doi.org/10.7930/J0862DC7

7. EPA, 2017: Multi-model Framework for Quantitative Sectoral Impacts Analysis: A Technical Report for the Fourth National Climate Assessment. EPA 430-R-17-001. U.S. Environmental Protection Agency (EPA), Washington, DC, 271 pp. https://cfpub.epa.gov/si/si_public_record_Report.cfm?dirEntryId=335095

8. Melillo, J.M., T.C. Richmond, and G.W. Yohe, Eds., 2014: *Highlights of Climate Change Impacts in the United States: The Third National Climate Assessment*. U.S. Global Change Research Program, Washington, DC, 148 pp. http://dx.doi.org/10.7930/J0H41PB6

9. van Vuuren, D.P., J. Edmonds, M. Kainuma, K. Riahi, A. Thomson, K. Hibbard, G.C. Hurtt, T. Kram, V. Krey, and J.F. Lamarque, 2011: The representative concentration pathways: An overview. *Climatic Change*, **109** (1-2), 5-31. http://dx.doi.org/10.1007/s10584-011-0148-z

10. Hayhoe, K., J. Edmonds, R.E. Kopp, A.N. LeGrande, B.M. Sanderson, M.F. Wehner, and D.J. Wuebbles, 2017: Climate models, scenarios, and projections. *Climate Science Special Report: Fourth National Climate Assessment, Volume I*. Wuebbles, D.J., D.W. Fahey, K.A. Hibbard, D.J. Dokken, B.C. Stewart, and T.K. Maycock, Eds. U.S. Global Change Research Program, Washington, DC, USA, 133-160. http://dx.doi.org/10.7930/J0WH2N54

11. National Academies of Sciences, Engineering, and Medicine, 2016: *Characterizing Risk in Climate Change Assessments: Proceedings of a Workshop*. Beatty, A., Ed. The National Academies Press, Washington, DC, 100 pp. http://dx.doi.org/10.17226/23569

12. Mastrandrea, M.D., C.B. Field, T.F. Stocker, O. Edenhofer, K.L. Ebi, D.J. Frame, H. Held, E. Kriegler, K.J. Mach, P.R. Matschoss, G.-K. Plattner, G.W. Yohe, and F.W. Zwiers, 2010: Guidance Note for Lead Authors of the IPCC Fifth Assessment Report on Consistent Treatment of Uncertainties. Intergovernmental Panel on Climate Change (IPCC), 7 pp. https://www.ipcc.ch/pdf/supporting-material/uncertainty-guidance-note.pdf

Summary Findings

NCA4 Summary Findings

These Summary Findings represent a high-level synthesis of the material in the underlying report. The findings consolidate Key Messages and supporting evidence from 16 national-level topic chapters, 10 regional chapters, and 2 chapters that focus on societal response strategies (mitigation and adaptation). Unless otherwise noted, qualitative statements regarding future conditions in these Summary Findings are broadly applicable across the range of different levels of future climate change and associated impacts considered in this report.

1. Communities

Climate change creates new risks and exacerbates existing vulnerabilities in communities across the United States, presenting growing challenges to human health and safety, quality of life, and the rate of economic growth.

The impacts of climate change are already being felt in communities across the country. More frequent and intense extreme weather and climate-related events, as well as changes in average climate conditions, are expected to continue to damage infrastructure, ecosystems, and social systems that provide essential benefits to communities. Future climate change is expected to further disrupt many areas of life, exacerbating existing challenges to prosperity posed by aging and deteriorating infrastructure, stressed ecosystems, and economic inequality. Impacts within and across regions will not be distributed equally. People who are already vulnerable, including lower-income and other marginalized communities, have lower capacity to prepare for and cope with extreme weather and climate-related events and are expected to experience greater impacts. Prioritizing adaptation actions for the most vulnerable populations would contribute to a more equitable future within and across communities. Global action to significantly cut greenhouse gas emissions can substantially reduce climate-related risks and increase opportunities for these populations in the longer term.

2. Economy

Without substantial and sustained global mitigation and regional adaptation efforts, climate change is expected to cause growing losses to American infrastructure and property and impede the rate of economic growth over this century.

In the absence of significant global mitigation action and regional adaptation efforts, rising temperatures, sea level rise, and changes in extreme events are expected to increasingly disrupt and damage critical infrastructure and property, labor productivity, and the vitality of our communities. Regional economies and industries that depend on natural resources and favorable climate conditions, such as agriculture, tourism, and fisheries, are vulnerable to the growing impacts of climate change. Rising temperatures are projected to reduce the efficiency of power generation while increasing energy demands, resulting in higher electricity costs. The impacts of climate change beyond our borders are expected to increasingly affect our trade and economy, including import and export prices and U.S. businesses with overseas operations and supply chains. Some aspects of our economy may see slight near-term improvements in a modestly warmer world. However, the continued warming that is projected to occur without substantial and sustained reductions in global greenhouse gas emissions is expected to cause substantial net damage to the U.S. economy throughout this century, especially in the absence of increased adaptation efforts. With continued growth in emissions at historic rates, annual losses in some economic sectors are projected to reach hundreds of billions of dollars by the end of the century—more than the current gross domestic product (GDP) of many U.S. states.

3. Interconnected Impacts

Climate change affects the natural, built, and social systems we rely on individually and through their connections to one another. These interconnected systems are increasingly vulnerable to cascading impacts that are often difficult to predict, threatening essential services within and beyond the Nation's borders.

Climate change presents added risks to interconnected systems that are already exposed to a range of stressors such as aging and deteriorating infrastructure, land-use changes, and population growth. Extreme weather and climate-related impacts on one system can result in increased risks or failures in other critical systems, including water resources, food production and distribution, energy and transportation, public health, international trade, and national security. The full extent of climate change risks to interconnected systems, many of which span regional and national boundaries, is often greater than the sum of risks to individual sectors. Failure to anticipate interconnected impacts can lead to missed opportunities for effectively managing the risks of climate change and can also lead to management responses that increase risks to other sectors and regions. Joint planning with stakeholders across sectors, regions, and jurisdictions can help identify critical risks arising from interaction among systems ahead of time.

4. Actions to Reduce Risks

Communities, governments, and businesses are working to reduce risks from and costs associated with climate change by taking action to lower greenhouse gas emissions and implement adaptation strategies. While mitigation and adaptation efforts have expanded substantially in the last four years, they do not yet approach the scale considered necessary to avoid substantial damages to the economy, environment, and human health over the coming decades.

Future risks from climate change depend primarily on decisions made today. The integration of climate risk into decision-making and the implementation of adaptation activities have significantly increased since the Third National Climate Assessment in 2014, including in areas of financial risk reporting, capital investment planning, development of engineering standards, military planning, and disaster risk management. Transformations in the energy sector—including the displacement of coal by natural gas and increased deployment of renewable energy—along with policy actions at the national, regional, state, and local levels are reducing greenhouse gas emissions in the United States. While these adaptation and mitigation measures can help reduce damages in a number of sectors, this assessment shows that more immediate and substantial global greenhouse gas emissions reductions, as well as regional adaptation efforts, would be needed to avoid the most severe consequences in the long term. Mitigation and adaptation actions also present opportunities for additional benefits that are often more immediate and localized, such as improving local air quality and economies through investments in infrastructure. Some benefits, such as restoring ecosystems and increasing community vitality, may be harder to quantify.

5. Water

The quality and quantity of water available for use by people and ecosystems across the country are being affected by climate change, increasing risks and costs to agriculture, energy production, industry, recreation, and the environment.

Rising air and water temperatures and changes in precipitation are intensifying droughts, increasing heavy downpours, reducing snowpack, and causing declines in surface water quality, with varying impacts across regions. Future warming will add to the stress on water supplies and adversely impact the availability of water in parts of the United States. Changes in the relative amounts and timing of snow and rainfall are leading to mismatches between water availability and needs in some regions, posing threats to, for example, the future reliability of hydropower production in the Southwest and the Northwest. Groundwater depletion is exacerbating drought risk in many parts of the United States, particularly in the Southwest and Southern Great Plains. Dependable and safe water supplies for U.S. Caribbean, Hawai'i, and U.S.-Affiliated Pacific Island communities are threatened by drought, flooding, and saltwater contamination due to sea level rise. Most U.S. power plants rely on a steady supply of water for cooling, and operations are expected to be affected by changes in water availability and temperature increases. Aging and deteriorating water infrastructure, typically designed for past environmental conditions, compounds the climate risk faced by society. Water management strategies that account for changing climate conditions can help reduce present and future risks to water security, but implementation of such practices remains limited.

6. Health

Impacts from climate change on extreme weather and climate-related events, air quality, and the transmission of disease through insects and pests, food, and water increasingly threaten the health and well-being of the American people, particularly populations that are already vulnerable.

Changes in temperature and precipitation are increasing air quality and health risks from wildfire and ground-level ozone pollution. Rising air and water temperatures and more intense extreme events are expected to increase exposure to waterborne and foodborne diseases, affecting food and water safety. With continued warming, cold-related deaths are projected to decrease and heat-related deaths are projected to increase; in most regions, increases in heat-related deaths are expected to outpace reductions in cold-related deaths. The frequency and severity of allergic illnesses, including asthma and hay fever, are expected to increase as a result of a changing climate. Climate change is also projected to alter the geographic range and distribution of disease-carrying insects and pests, exposing more people to ticks that carry Lyme disease and mosquitoes that transmit viruses such as Zika, West Nile, and dengue, with varying impacts across regions. Communities in the Southeast, for example, are particularly vulnerable to the combined health impacts from vector-borne disease, heat, and flooding. Extreme weather and climate-related events can have lasting mental health consequences in affected communities, particularly if they result in degradation of livelihoods or community relocation. Populations including older adults, children, low-income communities, and some communities of color are often disproportionately affected by, and less resilient to, the health impacts of climate change. Adaptation and mitigation policies and programs that help individuals, communities, and states prepare for the risks of a changing climate reduce the number of injuries, illnesses, and deaths from climate-related health outcomes.

7. Indigenous Peoples

Climate change increasingly threatens Indigenous communities' livelihoods, economies, health, and cultural identities by disrupting interconnected social, physical, and ecological systems.

Many Indigenous peoples are reliant on natural resources for their economic, cultural, and physical well-being and are often uniquely affected by climate change. The impacts of climate change on water, land, coastal areas, and other natural resources, as well as infrastructure and related services, are expected to increasingly disrupt Indigenous peoples' livelihoods and economies, including agriculture and agroforestry, fishing, recreation, and tourism. Adverse impacts on subsistence activities have already been observed. As climate changes continue, adverse impacts on culturally significant species and resources are expected to result in negative physical and mental health effects. Throughout the United States, climate-related impacts are causing some Indigenous peoples to consider or actively pursue community relocation as an adaptation strategy, presenting challenges associated with maintaining cultural and community continuity. While economic, political, and infrastructure limitations may affect these communities' ability to adapt, tightly knit social and cultural networks present opportunities to build community capacity and increase resilience. Many Indigenous peoples are taking steps to adapt to climate change impacts structured around self-determination and traditional knowledge, and some tribes are pursuing mitigation actions through development of renewable energy on tribal lands.

8. Ecosystems and Ecosystem Services

Ecosystems and the benefits they provide to society are being altered by climate change, and these impacts are projected to continue. Without substantial and sustained reductions in global greenhouse gas emissions, transformative impacts on some ecosystems will occur; some coral reef and sea ice ecosystems are already experiencing such transformational changes.

Many benefits provided by ecosystems and the environment, such as clean air and water, protection from coastal flooding, wood and fiber, crop pollination, hunting and fishing, tourism, cultural identities, and more will continue to be degraded by the impacts of climate change. Increasing wildfire frequency, changes in insect and disease outbreaks, and other stressors are expected to decrease the ability of U.S. forests to support economic activity, recreation, and subsistence activities. Climate change has already had observable impacts on biodiversity, ecosystems, and the benefits they provide to society. These impacts include the migration of native species to new areas and the spread of invasive species. Such changes are projected to continue, and without substantial and sustained reductions in global greenhouse gas emissions, extinctions and transformative

impacts on some ecosystems cannot be avoided in the long term. Valued aspects of regional heritage and quality of life tied to ecosystems, wildlife, and outdoor recreation will change with the climate, and as a result, future generations can expect to experience and interact with the natural environment in ways that are different from today. Adaptation strategies, including prescribed burning to reduce fuel for wildfire, creation of safe havens for important species, and control of invasive species, are being implemented to address emerging impacts of climate change. While some targeted response actions are underway, many impacts, including losses of unique coral reef and sea ice ecosystems, can only be avoided by significantly reducing global emissions of carbon dioxide and other greenhouse gases.

9. Agriculture and Food

Rising temperatures, extreme heat, drought, wildfire on rangelands, and heavy downpours are expected to increasingly disrupt agricultural productivity in the United States. Expected increases in challenges to livestock health, declines in crop yields and quality, and changes in extreme events in the United States and abroad threaten rural livelihoods, sustainable food security, and price stability.

Climate change presents numerous challenges to sustaining and enhancing crop productivity, livestock health, and the economic vitality of rural communities. While some regions (such as the Northern Great Plains) may see conditions conducive to expanded or alternative crop productivity over the next few decades, overall, yields from major U.S. crops are expected to decline as a consequence of increases in

temperatures and possibly changes in water availability, soil erosion, and disease and pest outbreaks. Increases in temperatures during the growing season in the Midwest are projected to be the largest contributing factor to declines in the productivity of U.S. agriculture. Projected increases in extreme heat conditions are expected to lead to further heat stress for livestock, which can result in large economic losses for producers. Climate change is also expected to lead to large-scale shifts in the availability and prices of many agricultural products across the world, with corresponding impacts on U.S. agricultural producers and the U.S. economy. These changes threaten future gains in commodity crop production and put rural livelihoods at risk. Numerous adaptation strategies are available to cope with adverse impacts of climate variability and change on agricultural production. These include altering what is produced, modifying the inputs used for production, adopting new technologies, and adjusting management strategies. However, these strategies have limits under severe climate change impacts and would require sufficient long- and short-term investment in changing practices.

10. Infrastructure

Our Nation's aging and deteriorating infrastructure is further stressed by increases in heavy precipitation events, coastal flooding, heat, wildfires, and other extreme events, as well as changes to average precipitation and temperature. Without adaptation, climate change will continue to degrade infrastructure performance over the rest of the century, with the potential for cascading impacts that threaten our economy, national security, essential services, and health and well-being.

Climate change and extreme weather events are expected to increasingly disrupt our Nation's energy and transportation systems, threatening more frequent and longer-lasting power outages, fuel shortages, and service disruptions, with cascading impacts on other critical sectors. Infrastructure currently designed for historical climate conditions is more vulnerable to future weather extremes and climate change. The continued increase in the frequency and extent of high-tide flooding due to sea level rise threatens America's trillion-dollar coastal property market and public infrastructure, with cascading impacts to the larger economy. In Alaska, rising temperatures and erosion are causing damage to buildings and coastal infrastructure that will be costly to repair or replace, particularly in rural areas; these impacts are expected to grow without adaptation. Expected increases in the severity and frequency of heavy precipitation events will affect inland infrastructure in every region, including access to roads, the viability of bridges, and the safety of pipelines. Flooding from heavy rainfall, storm surge, and rising high tides is expected to compound existing issues with aging infrastructure in the Northeast. Increased drought risk will threaten oil and gas drilling and refining, as well as electricity generation from power plants that rely on surface water for cooling. Forward-looking infrastructure design, planning, and operational measures and standards can reduce exposure and vulnerability to the impacts of climate change and reduce energy use while providing additional near-term benefits, including reductions in greenhouse gas emissions.

11. Oceans and Coasts

Coastal communities and the ecosystems that support them are increasingly threatened by the impacts of climate change. Without significant reductions in global greenhouse gas emissions and regional adaptation measures, many coastal regions will be transformed by the latter part of this century, with impacts affecting other regions and sectors. Even in a future with lower greenhouse gas emissions, many communities are expected to suffer financial impacts as chronic high-tide flooding leads to higher costs and lower property values.

Rising water temperatures, ocean acidification, retreating arctic sea ice, sea level rise, high-tide flooding, coastal erosion, higher storm surge, and heavier precipitation events threaten our oceans and coasts. These effects are projected to continue, putting ocean and marine species at risk, decreasing the productivity of certain fisheries, and threatening communities that rely on marine ecosystems for livelihoods and recreation, with particular impacts on fishing communities in Hawai'i and the U.S.-Affiliated Pacific Islands, the U.S. Caribbean, and the Gulf of Mexico. Lasting damage to coastal property and infrastructure driven by sea level rise and storm surge is expected to lead to financial losses for individuals, businesses, and communities, with the Atlantic and Gulf Coasts facing above-average risks. Impacts on coastal energy and transportation infrastructure driven by sea level rise and storm surge have the potential for cascading costs and disruptions across the country. Even if significant emissions reductions occur, many of the effects from sea level rise over this century—and particularly through mid-century—are already locked in due to historical emissions, and many communities are already dealing with the consequences. Actions to plan for and adapt to more frequent, widespread, and severe coastal flooding, such as shoreline protection and conservation of coastal ecosystems, would decrease direct losses and cascading impacts on other sectors and parts of the country. More than half of the damages to coastal property are estimated to be avoidable through well-timed adaptation measures. Substantial and sustained reductions in global greenhouse gas emissions would also significantly reduce projected risks to fisheries and communities that rely on them.

12. Tourism and Recreation

Outdoor recreation, tourist economies, and quality of life are reliant on benefits provided by our natural environment that will be degraded by the impacts of climate change in many ways.

Climate change poses risks to seasonal and outdoor economies in communities across the United States, including impacts on economies centered around coral reef-based recreation, winter recreation, and inland water-based recreation. In turn, this affects the well-being of the people who make their living supporting these economies, including rural, coastal, and Indigenous communities. Projected increases in wildfire smoke events are expected to impair outdoor recreational activities and visibility in wilderness areas. Declines in snow and ice cover caused by warmer winter temperatures are expected to negatively impact the winter recreation industry in the Northwest, Northern Great Plains, and the Northeast. Some fish, birds, and mammals are expected to shift where they live as a result of climate change, with implications for hunting, fishing, and other wildlife-related activities. These and other climate-related impacts are expected to result in decreased tourism revenue in some places and, for some communities, loss of identity. While some new opportunities may emerge from these ecosystem changes, cultural identities and economic and recreational opportunities based around historical use of and interaction with species or natural resources in many areas are at risk. Proactive management strategies, such as the use of projected stream temperatures to set priorities for fish conservation, can help reduce disruptions to tourist economies and recreation.

20

Overview

22

1 Overview

Howe Ridge Fire in Montana's Glacier National Park on August 12, 2018. *Photo credit: National Park Service.*

Federal Coordinating Lead Author
David Reidmiller, U.S. Global Change Research Program

Chapter Lead
Alexa Jay, U.S. Global Change Research Program

Chapter Authors
Christopher W. Avery, U.S. Global Change Research Program
Daniel Barrie, National Oceanic and Atmospheric Administration
Apurva Dave, U.S. Global Change Research Program
Benjamin DeAngelo, National Oceanic and Atmospheric Administration
Matthew Dzaugis, U.S. Global Change Research Program
Michael Kolian, U.S. Environmental Protection Agency
Kristin Lewis, U.S. Global Change Research Program
Katie Reeves, U.S. Global Change Research Program
Darrell Winner, U.S. Environmental Protection Agency

Recommended Citation for Chapter

Jay, A., D.R. Reidmiller, C.W. Avery, D. Barrie, B.J. DeAngelo, A. Dave, M. Dzaugis, M. Kolian, K.L.M. Lewis, K. Reeves, and D. Winner, 2018: Overview. In *Impacts, Risks, and Adaptation in the United States: Fourth National Climate Assessment, Volume II* [Reidmiller, D.R., C.W. Avery, D.R. Easterling, K.E. Kunkel, K.L.M. Lewis, T.K. Maycock, and B.C. Stewart (eds.)]. U.S. Global Change Research Program, Washington, DC, USA. doi: 10.7930/NCA4.2018.CH1

Introduction

Earth's climate is now changing faster than at any point in the history of modern civilization, primarily as a result of human activities. The impacts of global climate change are already being felt in the United States and are projected to intensify in the future—but the severity of future impacts will depend largely on actions taken to reduce greenhouse gas emissions and to adapt to the changes that will occur. Americans increasingly recognize the risks climate change poses to their everyday lives and livelihoods and are beginning to respond (Figure 1.1). Water managers in the Colorado River Basin have mobilized users to conserve water in response to ongoing drought intensified by higher temperatures, and an extension program in Nebraska is helping ranchers reduce drought and heat risks to their operations. The state of Hawai'i is developing management options to promote coral reef recovery from widespread bleaching events caused by warmer waters that threaten tourism, fisheries, and coastal protection from wind and waves. To address higher risks of flooding from heavy rainfall, local governments in southern Louisiana are pooling hazard reduction funds, and cities and states in the Northeast are investing in more resilient water, energy, and transportation infrastructure. In Alaska, a tribal health organization is developing adaptation strategies to address physical and mental health challenges driven by climate change and other environmental changes. As Midwestern farmers adopt new management strategies to reduce erosion and nutrient losses caused by heavier rains, forest managers in the Northwest are developing adaptation strategies in response to wildfire increases that affect human health, water resources, timber production, fish and wildlife, and recreation. After extensive hurricane damage fueled in part by a warmer atmosphere and warmer, higher seas, communities in Texas are considering ways to rebuild more resilient infrastructure. In the U.S. Caribbean, governments are developing new frameworks for storm recovery based on lessons learned from the 2017 hurricane season.

Climate-related risks will continue to grow without additional action. Decisions made today determine risk exposure for current and future generations and will either broaden or limit options to reduce the negative consequences of climate change. While Americans are responding in ways that can bolster resilience and improve livelihoods, neither global efforts to mitigate the causes of climate change nor regional efforts to adapt to the impacts currently approach the scales needed to avoid substantial damages to the U.S. economy, environment, and human health and well-being over the coming decades.

Americans Respond to the Impacts of Climate Change

Alaska

Impact	Action
The physical and mental health of rural Alaskans is increasingly challenged by unpredictable weather and other environmental changes.	The Alaska Native Tribal Health Consortium's Center for Climate and Health is using novel adaptation strategies to reduce climate-related risk including difficulty in harvesting local foods and more hazardous travel conditions.

Northern Great Plains

Impact	Action
Flash droughts and extreme heat illustrate sustainability challenges for ranching operations, with emergent impacts on rural prosperity and mental health.	The National Drought Mitigation Center is helping ranchers plan to reduce drought and heat risks to their operations.

Midwest

Impact	Action
Increasing heavy rains are leading to more soil erosion and nutrient loss on Midwestern cropland.	Iowa State developed a program using prairie strips in farm fields to reduce soil and nutrient loss while increasing biodiversity.

Northwest

Impact
Wildfire increases and associated smoke are affecting human health, water resources, timber production, fish and wildlife and recreation.

Action
Federal forests have developed adaptation strategies for climate change that include methods to address increasing wildfire risks.

Northeast

Impact
Water, energy, and transportation infrastructure are affected by snow storms, drought, heat waves, and flooding.

Action
Cities and states throughout the region are assessing their vulnerability to climate change and making investments to increase infrastructure resilience.

Southwest

Impact	Action
Drought in the Colorado River basin reduced Lake Mead by over half since 2000, increasing risk of water shortages for cities, farms, and ecosystems.	Seven U.S. state governments and U.S. and Mexico federal governments mobilized users to converse water, keeping the lake above a critical level.

Southern Great Plains

Impact	Action
Hurricane Harvey's landfall on the Texas coast in 2017 was one of the costliest natural disasters in U.S. history.	The Governor's Commission to Rebuild Texas was created to support the economic recovery and rebuilding of infrastructure in affected Texas communities.

Southeast

Impact	Action
Flooding in Louisiana is increasing from extreme rainfall.	The Acadiana Planning Commission in Louisiana is pooling hazard reduction funds to address increasing flood risk.

Hawai'i and U.S. -Affiliated Pacific Islands

Impact	Action
The 2015 coral bleaching event resulted in an average mortality of 50% of the coral cover in western Hawai'i alone.	A state working group generated management options to promote recovery and reduce threats to coral reefs.

U.S. Caribbean

Impact	Action
Damages from the 2017 hurricanes have been compounded by slow recovery of energy, communications, and transportation systems, impacting all social and economic sectors.	The U.S. Virgin Islands Governor's Office led a workshop aimed at gathering lessons from the initial hurricane response and establishing a framework for recovery and resilience.

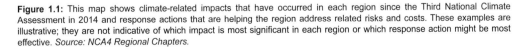

Figure 1.1: This map shows climate-related impacts that have occurred in each region since the Third National Climate Assessment in 2014 and response actions that are helping the region address related risks and costs. These examples are illustrative; they are not indicative of which impact is most significant in each region or which response action might be most effective. *Source: NCA4 Regional Chapters.*

Climate shapes where and how we live and the environment around us. Natural ecosystems, agricultural systems, water resources, and the benefits they provide to society are adapted to past climate conditions and their natural range of variability. A water manager may use past or current streamflow records to design a dam, a city could issue permits for coastal development based on current flood maps, and an electric utility or a farmer may invest in equipment suited to the current climate, all with the expectation that their investments and management practices will meet future needs.

However, the assumption that current and future climate conditions will resemble the recent past is no longer valid (Ch. 28: Adaptation, KM 2). Observations collected around the world provide significant, clear, and compelling evidence that global average temperature is much higher, and is rising more rapidly, than anything modern civilization has experienced, with widespread and growing impacts (Figure 1.2) (CSSR, Ch. 1.9). The warming trend observed over the past century can only be explained by the effects that human activities, especially emissions of greenhouse gases, have had on the climate (Ch. 2: Climate, KM 1 and Figure 2.1).

Climate change is transforming where and how we live and presents growing challenges to human health and quality of life, the economy, and the natural systems that support us. Risks posed by climate variability and change vary by region and sector and by the vulnerability of people experiencing impacts. Social, economic, and geographic factors shape the exposure of people and communities to climate-related impacts and their capacity to respond. Risks are often highest for those that are already vulnerable, including low-income communities, some communities of color, children, and the elderly (Ch. 14: Human Health, KM 2; Ch. 15: Tribes, KM 1–3; Ch. 28: Adaptation, Introduction). Climate change threatens to exacerbate existing social and economic inequalities that result in higher exposure and sensitivity to extreme weather and climate-related events and other changes (Ch. 11: Urban, KM 1). Marginalized populations may also be affected disproportionately by actions to address the underlying causes and impacts of climate change, if they are not implemented under policies that consider existing inequalities (Ch. 11: Urban, KM 4; Ch. 28: Adaptation, KM 4).

This report draws a direct connection between the warming atmosphere and the resulting changes that affect Americans' lives, communities, and livelihoods, now and in the future. It documents vulnerabilities, risks, and impacts associated with natural climate variability and human-caused climate change across the United States and provides examples of response actions underway in many communities. It concludes that *the evidence of human-caused climate change is overwhelming and continues to strengthen, that the impacts of climate change are intensifying across the country, and that climate-related threats to Americans' physical, social, and economic well-being are rising.* These impacts are projected to intensify—but how much they intensify will depend on actions taken to reduce global greenhouse gas emissions and to adapt to the risks from climate change now and in the coming decades (Ch. 28: Adaptation, Introduction; Ch. 29: Mitigation, KM 3 and 4).

Our Changing Climate: Observations, Causes, and Future Change

Observed Change

Observations from around the world show the widespread effects of increasing greenhouse gas concentrations on Earth's climate. High temperature extremes and heavy precipitation events are increasing. Glaciers and snow cover are shrinking, and sea ice is retreating. Seas are warming, rising, and becoming more acidic, and marine species are moving to new locations toward cooler waters. Flooding is becoming more frequent along the U.S. coastline. Growing seasons are lengthening, and wildfires are increasing. These and many other changes are clear signs of a warming world (Figure 1.2) (Ch. 2: Climate, Box 2.2; App. 3: Data & Scenarios, see also the USGCRP Indicators and EPA Indicators websites).

California Drought Affects Mountain Snowpack
California's recent multiyear drought left Tioga Pass in the Sierra Nevada mountain range nearly snowless at the height of winter in January 2015. *Photo credit: Bartshé Miller.*

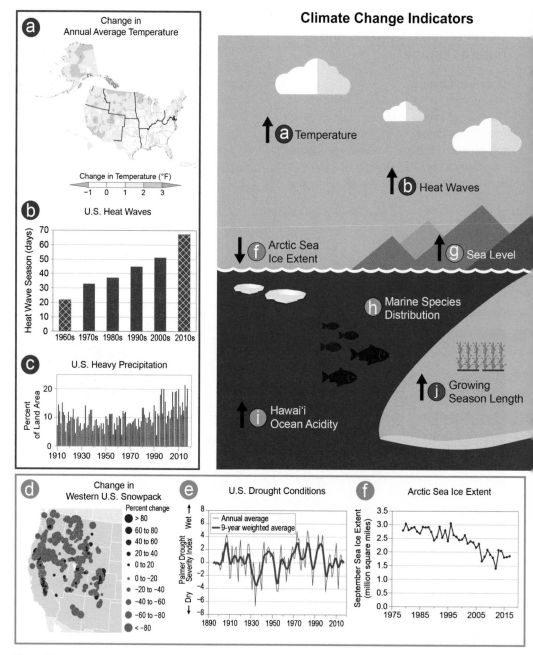

Figure 1.2: Long-term observations demonstrate the warming trend in the climate system and the effects of increasing atmospheric greenhouse gas concentrations (Ch. 2: Climate, Box 2.2). This figure shows climate-relevant indicators of change

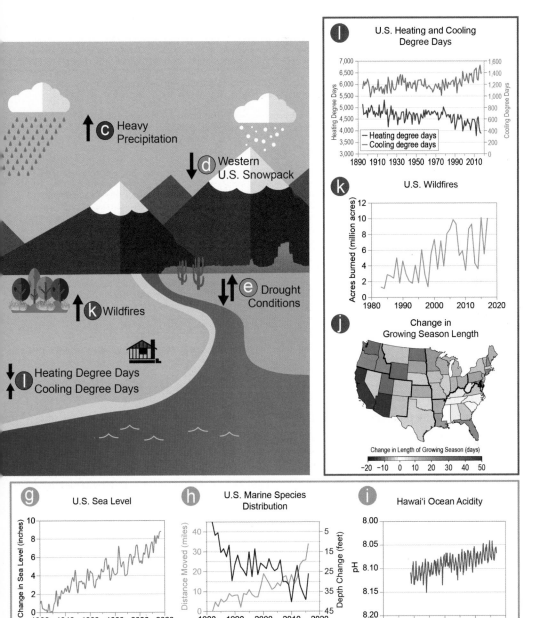

based on data collected across the United States. Upward-pointing arrows indicate an increasing trend; downward-pointing arrows indicate a decreasing trend. Bidirectional arrows (e.g., for drought conditions) indicate a lack of a definitive national trend.

(Figure caption continued on next page)

Atmosphere (a–c): (a) Annual average temperatures have increased by 1.8°F across the contiguous United States since the beginning of the 20th century; this figure shows observed change for 1986–2016 (relative to 1901–1960 for the contiguous United States and 1925–1960 for Alaska, Hawai'i, Puerto Rico, and the U.S. Virgin Islands). Alaska is warming faster than any other state and has warmed twice as fast as the global average since the mid-20th century (Ch. 2: Climate, KM 5; Ch. 26: Alaska, Background). (b) The season length of heat waves in many U.S. cities has increased by over 40 days since the 1960s. Hatched bars indicate partially complete decadal data. (c) The relative amount of annual rainfall that comes from large, single-day precipitation events has changed over the past century; since 1910, a larger percentage of land area in the contiguous United States receives precipitation in the form of these intense single-day events.

Ice, snow, and water (d–f): (d) Large declines in snowpack in the western United States occurred from 1955 to 2016. (e) While there are a number of ways to measure drought, there is currently no detectable change in long-term U.S. drought statistics using the Palmer Drought Severity Index. (f) Since the early 1980s, the annual minimum sea ice extent (observed in September each year) in the Arctic Ocean has decreased at a rate of 11%–16% per decade (Ch. 2: Climate, KM 7).

Oceans and coasts (g–i): (g) Annual median sea level along the U.S. coast (with land motion removed) has increased by about 9 inches since the early 20th century as oceans have warmed and land ice has melted (Ch. 2: Climate, KM 4). (h) Fish, shellfish, and other marine species along the Northeast coast and in the eastern Bering Sea have, on average, moved northward and to greater depths toward cooler waters since the early 1980s (records start in 1982). (i) Oceans are also currently absorbing more than a quarter of the carbon dioxide emitted to the atmosphere annually by human activities, increasing their acidity (measured by lower pH values; Ch. 2: Climate, KM 3).

Land and ecosystems (j–l): (j) The average length of the growing season has increased across the contiguous United States since the early 20th century, meaning that, on average, the last spring frost occurs earlier and the first fall frost arrives later; this map shows changes in growing season length at the state level from 1895 to 2016. (k) Warmer and drier conditions have contributed to an increase in large forest fires in the western United States and Interior Alaska over the past several decades (CSSR, Ch. 8.3). (l) Degree days are defined as the number of degrees by which the average daily temperature is higher than 65°F (cooling degree days) or lower than 65°F (heating degree days) and are used as a proxy for energy demands for cooling or heating buildings. Changes in temperatures indicate that heating needs have decreased and cooling needs have increased in the contiguous United States over the past century.

Sources: (a) adapted from Vose et al. 2017, (b) EPA, (c–f and h–l) adapted from EPA 2016, (g and center infographic) EPA and NOAA.

Causes of Change

Scientists have understood the fundamental physics of climate change for almost 200 years. In the 1850s, researchers demonstrated that carbon dioxide and other naturally occurring greenhouse gases in the atmosphere prevent some of the heat radiating from Earth's surface from escaping to space: this is known as the greenhouse effect. This natural greenhouse effect warms the planet's surface about 60°F above what it would be otherwise, creating a habitat suitable for life. Since the late 19th century, however, humans have released an increasing amount of greenhouse gases into the atmosphere through burning fossil fuels and, to a lesser extent, deforestation and land-use change. As a result, the atmospheric concentration of carbon dioxide, the largest contributor to human-caused warming, has increased by about 40% over the industrial era. This change has intensified the natural greenhouse effect, driving an increase in global surface temperatures and other widespread changes in Earth's climate that are unprecedented in the history of modern civilization.

Global climate is also influenced by natural factors that determine how much of the sun's energy enters and leaves Earth's atmosphere and by natural climate cycles that affect temperatures and weather patterns in the short term, especially regionally (see Ch. 2: Climate, Box 2.1). However, the unambiguous long-term warming trend in global average temperature over the last century cannot be explained by natural factors alone. Greenhouse gas emissions from human activities are the only factors that can account for the observed warming over the last century; there are no credible alternative human or natural explanations supported by the observational evidence. Without human activities, the influence of natural factors alone would actually have had a slight cooling effect on global climate over the last 50 years (Ch. 2: Climate, KM 1, Figure 2.1).

Future Change

Greenhouse gas emissions from human activities will continue to affect Earth's climate for decades and even centuries. Humans are adding carbon dioxide to the atmosphere at a rate far greater than it is removed by natural processes, creating a long-lived reservoir of the gas in the atmosphere and oceans that is driving the climate to a warmer and warmer state. Some of the other greenhouse gases released by human activities, such as methane, are removed from the atmosphere by natural processes more quickly than carbon dioxide; as a result, efforts to cut emissions of these gases could help reduce the rate of global temperature increases over the next few decades. However, longer-term changes in climate will largely be determined by emissions and atmospheric concentrations of carbon dioxide and other longer-lived greenhouse gases (Ch. 2: Climate, KM 2).

Climate models representing our understanding of historical and current climate conditions are often used to project how our world will change under future conditions (see Ch. 2: Climate, Box 2.7). "Climate" is defined as weather conditions over multiple decades, and climate model projections are generally not designed to capture annual or even decadal variation in climate conditions. Instead, projections are typically used to capture long-term changes, such as how the climate system will respond

to changes in greenhouse gas levels over this century. Scientists test climate models by comparing them to current observations and historical changes. Confidence in these models is based, in part, on how well they reproduce these observed changes. Climate models have proven remarkably accurate in simulating the climate change we have experienced to date, particularly in the past 60 years or so when we have greater confidence in observations (see CSSR, Ch. 4.3.1). The observed signals of a changing climate continue to become stronger and clearer over time, giving scientists increased confidence in their findings even since the Third National Climate Assessment was released in 2014.

Today, the largest uncertainty in projecting future climate conditions is the level of greenhouse gas emissions going forward. Future global greenhouse gas emissions levels and resulting impacts depend on economic, political, and demographic factors that can be difficult to predict with confidence far into the future. Like previous climate assessments, NCA4 relies on a suite of possible scenarios to evaluate the implications of different climate outcomes and associated impacts throughout the 21st century. These "Representative Concentration Pathways" (RCPs) capture a range of potential greenhouse gas emissions pathways and associated atmospheric concentration levels through 2100.

RCPs drive climate model projections for temperature, precipitation, sea level, and other variables under futures that have either lower or higher greenhouse gas emissions. RCPs are numbered according to changes in radiative forcing by 2100 relative to preindustrial conditions: +2.6, +4.5, +6.0, or +8.5 watts per square meter (W/m²). Each RCP leads to a different

Box 1.1: Confidence and Uncertainty in Climate Science

Many of the decisions we make every day are based on less-than-perfect knowledge. For example, while GPS-based applications on smartphones can provide a travel-time estimate for our daily drive to work, an unexpected factor like a sudden downpour or fender bender might mean a ride originally estimated to be 20 minutes could actually take longer. Fortunately, even with this uncertainty we are confident that our trip is unlikely to take less than 20 minutes or more than half an hour—and we know where we are headed. We have enough information to plan our commute.

Uncertainty is also a part of science. A key goal of scientific research is to increase our confidence and reduce the uncertainty in our understanding of the world around us. Even so, there is no expectation that uncertainty can be fully eliminated, just as we do not expect a perfectly accurate estimate for our drive time each day. Studying Earth's climate system is particularly challenging because it integrates many aspects of a complex natural system as well as many human-made systems. Climate scientists find varying ranges of uncertainty in many areas, including observations of climate variables, the analysis and interpretation of those measurements, the development of new observational instruments, and the use of computer-based models of the processes governing Earth's climate system. While there is inherent uncertainty in climate science, there is high confidence in our understanding of the greenhouse effect and the knowledge that human activities are changing the climate in unprecedented ways. There is enough information to make decisions based on that understanding.

Where important uncertainties do exist, efforts to quantify and report those uncertainties can help decision-makers plan for a range of possible future outcomes. These efforts also help scientists advance understanding and ultimately increase confidence in and the usefulness of model projections. Assessments like this one explicitly address scientific uncertainty associated with findings and use specific language to express it to improve relevance to risk analysis and decision-making (see Front Matter and Box 1.2).

level of projected global temperature change; higher numbers indicate greater projected temperature change and associated impacts. The higher scenario (RCP8.5) represents a future where annual greenhouse gas emissions increase significantly throughout the 21st century before leveling off by 2100, whereas the other RCPs represent more rapid and substantial mitigation by mid-century, with greater reductions thereafter. Current trends in annual greenhouse gas emissions, globally, are consistent with RCP8.5.

Of the two RCPs predominantly referenced throughout this report, the lower scenario (RCP4.5) envisions about 85% lower greenhouse

gas emissions than the higher scenario (RCP8.5) by the end of the 21st century (see Ch. 2: Climate, Figure 2.2). In some cases, throughout this report, a very low scenario (RCP2.6) that represents more immediate, substantial, and sustained emissions reductions is considered. Each RCP could be consistent with a range of underlying socioeconomic conditions or policy choices. See the Scenario Products section of Appendix 3 in this report, as well as CSSR Chapters 4.2.1 and 10.2.1 for more detail.

The effects of different future greenhouse gas emissions levels on global climate become most evident around 2050, when temperature (Figure 1.3) (Ch. 2: Climate, Figure 2.2), precipitation,

Projected Changes in U.S. Annual Average Temperatures

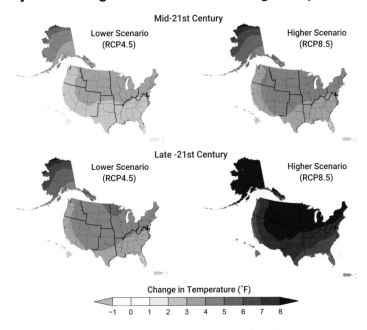

Figure 1.3: Annual average temperatures across the United States are projected to increase over this century, with greater changes at higher latitudes as compared to lower latitudes, and under a higher scenario (RCP8.5; right) than under a lower one (RCP4.5; left). This figure shows projected differences in annual average temperatures for mid-century (2036–2065; top) and end of century (2071–2100; bottom) relative to the near present (1986–2015). *From Figure 2.4, Ch. 2: Climate (Source: adapted from Vose et al. 2017).*

and sea level rise (Figure 1.4) (Ch. 2: Climate, Figure 2.3) projections based on each scenario begin to diverge significantly. With substantial and sustained reductions in greenhouse gas emissions (e.g., consistent with the very low scenario [RCP2.6]), the increase in global annual average temperature relative to preindustrial times could be limited to less than 3.6°F (2°C) (Ch. 2: Climate, Box 2.4; CSSR, Ch. 4.2.1). Without significant greenhouse gas mitigation, the increase in global annual average temperature could reach 9°F or more by the end of this century (Ch. 2: Climate, KM 2). For some aspects of Earth's climate system that take longer to respond to changes in atmospheric greenhouse gas concentrations, such as global sea level, some degree of long-term change will

be locked in for centuries to come, regardless of the future scenario (see CSSR, Ch. 12.5.3). Early greenhouse gas emissions mitigation can reduce climate impacts in the nearer term (such as reducing the loss of arctic sea ice and the effects on species that use it) and in the longer term by avoiding critical thresholds (such as marine ice sheet instability and the resulting consequences for global sea level and coastal development; Ch. 29: Mitigation, Timing and Magnitude of Action).

Annual average temperatures in the United States are projected to continue to increase in the coming decades. Regardless of future scenario, additional increases in temperatures across the contiguous United States of at least

Projected Relative Sea Level Change in the United States by 2100

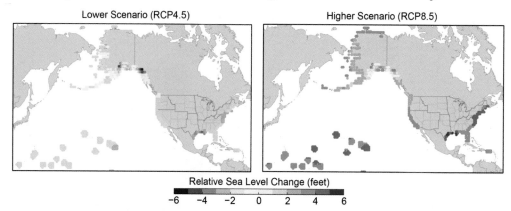

Figure 1.4: The maps show projections of change in relative sea level along the U.S. coast by 2100 (as compared to 2000) under the lower (RCP4.5) and higher (RCP8.5) scenarios (see CSSR, Ch. 12.5). Globally, sea levels will continue to rise from thermal expansion of the ocean and melting of land-based ice masses (such as Greenland, Antarctica, and mountain glaciers). Regionally, however, the amount of sea level rise will not be the same everywhere. Where land is sinking (as along the Gulf of Mexico coastline), relative sea level rise will be higher, and where land is rising (as in parts of Alaska), relative sea level rise will be lower. Changes in ocean circulation (such as the Gulf Stream) and gravity effects due to ice melt will also alter the heights of the ocean regionally. Sea levels are expected to continue to rise along almost all U.S. coastlines, and by 2100, under the higher scenario, coastal flood heights that today cause major damages to infrastructure would become common during high tides nationwide (Ch. 8: Coastal; Scenario Products section in Appendix 3). *Source: adapted from CSSR, Figure 12.4.*

2.3°F relative to 1986–2015 are expected by the middle of this century. As a result, recent record-setting hot years are expected to become common in the near future. By late this century, increases of 2.3°–6.7°F are expected under a lower scenario (RCP4.5) and 5.4°–11.0°F under a higher scenario (RCP8.5) relative to 1986–2015 (Figure 1.3) (Ch. 2: Climate, KM 5, Figure 2.4). Alaska has warmed twice as fast as the global average since the mid-20th century; this trend is expected to continue (Ch. 26: Alaska, Background).

High temperature extremes, heavy precipitation events, high tide flooding events along the U.S. coastline, ocean acidification and warming, and forest fires in the western United States and

Alaska are all projected to continue to increase, while land and sea ice cover, snowpack, and surface soil moisture are expected to continue to decline in the coming decades. These and other changes are expected to increasingly impact water resources, air quality, human health, agriculture, natural ecosystems, energy and transportation infrastructure, and many other natural and human systems that support communities across the country. The severity of these projected impacts, and the risks they present to society, is greater under futures with higher greenhouse gas emissions, especially if limited or no adaptation occurs (Ch. 29: Mitigation, KM 2).

Box 1.2: Evaluating Risks to Inform Decisions

In this report, *risks* are often defined in a qualitative sense as threats to life, health and safety, the environment, economic well-being, and other things of value to society (Ch. 28: Adaptation, Introduction). In some cases, risks are described in quantitative terms: estimates of how likely a given threat is to occur (probability) and the damages that would result if it did happen (consequences). Climate change is a risk management challenge for society; it presents uncertain—and potentially severe—consequences for natural and human systems across generations. It is characterized by multiple intersecting and uncertain future hazards and, therefore, acts as a risk multiplier that interacts with other stressors to create new risks or to alter existing ones (see Ch. 17: Complex Systems, KM 1).

Current and future greenhouse gas emissions, and thus mitigation actions to reduce emissions, will largely determine future climate change impacts and risks to society. Mitigation and adaptation activities can be considered complementary strategies—mitigation efforts can reduce future risks, while adaptation can minimize the consequences of changes that are already happening as a result of past and present greenhouse gas emissions. Adaptation entails proactive decision-making and investments by individuals, businesses, and governments to counter specific risks from climate change that vary from place to place. Climate risk management includes some familiar attributes and tactics for most businesses and local governments, which often manage or design for a variety of weather-related risks, including coastal and inland storms, heat waves, threats to water availability, droughts, and floods.

Measuring risk encompasses both likelihoods and consequences of specific outcomes and involves judgments about what is of value, ranking of priorities, and cost–benefit analyses that incorporate the tradeoffs among climate and non-climate related options. This report characterizes specific risks across regions and sectors in an effort to help people assess the risks they face, create and implement a response plan, and monitor and evaluate the efficacy of a given action (see Ch. 28: Adaptation, KM 1, Figure 28.1).

Climate Change in the United States: Current and Future Risks

Some climate-related impacts, such as increasing health risks from extreme heat, are common to many regions of the United States (Ch. 14: Human Health, KM 1). Others represent more localized risks, such as infrastructure damage caused by thawing of permafrost (long-frozen ground) in Alaska or threats to coral reef ecosystems from warmer and more acidic seas in the U.S. Caribbean, as well as Hawai'i and the U.S.-Affiliated Pacific Islands (Ch. 26: Alaska, KM 2; Ch. 20: U.S. Caribbean, KM 2; Ch. 27: Hawai'i & Pacific Islands, KM 4). Risks vary by both a community's exposure to physical climate impacts and by factors that influence its ability to respond to changing conditions and to recover from adverse weather and climate-related events such as extreme storms or wildfires (Ch. 14: Human Health, KM 2; Ch. 15: Tribes, State of the Sector, KM 1 and 2; Ch. 28: Adaptation, KM 4).

Many places are subject to more than one climate-related impact, such as extreme rainfall combined with coastal flooding, or drought coupled with extreme heat, wildfire, and flooding. The compounding effects of these impacts result in increased risks to people, infrastructure, and interconnected economic sectors (Ch. 11: Urban, KM 1). Impacts affecting

35

interconnected systems can cascade across sectors and regions, creating complex risks and management challenges. For example, changes in the frequency, intensity, extent, and duration of wildfires can result in a higher instance of landslides that disrupt transportation systems and the flow of goods and services within or across regions (Box 1.3). Many observed impacts reveal vulnerabilities in these interconnected systems that are expected to be exacerbated as climate-related risks intensify. Under a higher scenario (RCP8.5), it is very likely that some impacts, such as the effects of ice sheet disintegration on sea level rise and coastal development, will be irreversible for many thousands of years, and others, such as species extinction, will be permanent (Ch. 7: Ecosystems, KM 1; Ch. 9: Oceans, KM 1; Ch. 29: Mitigation, KM 2).

Economy and Infrastructure

Without more significant global greenhouse gas mitigation and regional adaptation efforts, climate change is expected to cause substantial losses to infrastructure and property and impede the rate of economic growth over this century (Ch. 4: Energy, KM 1; Ch. 8: Coastal, KM 1; Ch. 11: Urban, KM 2; Ch. 12: Transportation, KM 1; Regional Chapters 18–27). Regional economies and industries that depend on natural resources and favorable climate conditions, such as agriculture, tourism, and fisheries, are increasingly vulnerable to impacts driven by climate change (Ch. 7: Ecosystems, KM 3; Ch. 10: Agriculture, KM 1). Reliable and affordable energy supplies, which underpin virtually every sector of the economy, are increasingly at risk from climate change and weather extremes (Ch. 4: Energy, KM 1). The impacts of climate

Box 1.3: Interconnected Impacts of Climate Change

The impacts of climate change and extreme weather on natural and built systems are often considered from the perspective of individual sectors: how does a changing climate impact water resources, the electric grid, or the food system? None of these sectors, however, exists in isolation. The natural, built, and social systems we rely on are all interconnected, and impacts and management choices within one sector may have cascading effects on the others (Ch. 17: Complex Systems, KM 1).

For example, wildfire trends in the western United States are influenced by rising temperatures and changing precipitation patterns, pest populations, and land management practices. As humans have moved closer to forestlands, increased fire suppression practices have reduced natural fires and led to denser vegetation, resulting in fires that are larger and more damaging when they do occur (Figures 1.5 and 1.2k) (Ch. 6: Forests, KM 1). Warmer winters have led to increased pest outbreaks and significant tree kills, with varying feedbacks on wildfire. Increased wildfire driven by climate change is projected to increase costs associated with health effects, loss of homes and other property, wildfire response, and fuel management. Failure to anticipate these interconnected impacts can lead to missed opportunities for effectively managing risks within a single sector and may actually increase risks to other sectors. Planning around wildfire risk and other risks affected by climate change entails the challenge of accounting for all of these influences and how they interact with one another (see Ch. 17: Complex Systems, Box 17.4).

Box 1.3: Interconnected Impacts of Climate Change, *continued*

New to this edition of the NCA, Chapter 17 (Complex Systems) highlights several examples of interconnected impacts and documents how a multisector perspective and joint management of systems can enhance resilience to a changing climate. It is often difficult or impossible to quantify and predict how all relevant processes and interactions in interconnected systems will respond to climate change. Non-climate influences, such as population changes, add to the challenges of projecting future outcomes (Ch. 17: Complex Systems, KM 2). Despite these challenges, there are opportunities to learn from experience to guide future risk management decisions. Valuable lessons can be learned retrospectively: after Superstorm Sandy in 2012, for example, the mayor of New York City initiated a Climate Change Adaptation Task Force that brought together stakeholders from several sectors such as water, transportation, energy, and communications to address the interdependencies among them (Ch. 17: Complex Systems, Box 17.1, KM 3).

Wildfire at the Wildland–Urban Interface

Figure 1.5: Wildfires are increasingly encroaching on American communities, posing threats to lives, critical infrastructure, and property. In October 2017, more than a dozen fires burned through northern California, killing dozens of people and leaving thousands more homeless. Communities distant from the fires were affected by poor air quality as smoke plumes darkened skies and caused the cancellation of school and other activities across the region. (left) A NASA satellite image shows active fires on October 9, 2017. (right) The Tubbs Fire, which burned parts of Napa, Sonoma, and Lake counties, was the most destructive in California's history. It caused an estimated $1.2 billion in damages and destroyed over 5,000 structures, including 5% of the housing stock in the city of Santa Rosa. *Image credits: (left) NASA; (right) Master Sgt. David Loeffler, U.S. Air National Guard.*

change beyond our borders are expected to increasingly affect our trade and economy, including import and export prices and U.S. businesses with overseas operation and supply chains (Box 1.4) (Ch. 16: International, KM 1; Ch. 17: Complex Systems, KM 1). Some aspects of our economy may see slight improvements in a modestly warmer world. However, the continued warming that is projected to occur without significant reductions in global greenhouse gas emissions is expected to cause substantial net damage to the U.S. economy, especially in the absence of increased adaptation efforts. The potential for losses in some sectors could reach hundreds of billions of dollars per year by the end of this century (Ch. 29: Mitigation, KM 2).

Existing water, transportation, and energy infrastructure already face challenges from heavy rainfall, inland and coastal flooding, landslides, drought, wildfire, heat waves, and other weather and climate events (Figures 1.5–1.9)

(Ch. 11: Urban, KM 2; Ch. 12: Transportation, KM 1). Many extreme weather and climate-related events are expected to become more frequent and more intense in a warmer world, creating greater risks of infrastructure disruption and failure that can cascade across economic sectors (Ch. 3: Water, KM 2; Ch. 4: Energy, KM 1; Ch. 11: Urban, KM 3; Ch. 12: Transportation, KM 2). For example, more frequent and severe heat waves and other extreme events in many parts of the United States are expected to increase stresses on the energy system, amplifying the risk of more frequent and longer-lasting power outages and fuel shortages that could affect other critical sectors and systems, such as access to medical care (Ch. 17: Complex Systems, Box 17.5; Ch. 4: Energy, KM 1; Ch. 8: Coastal, KM 1; Ch. 11: Urban, KM 3; Ch. 12: Transportation, KM 3). Current infrastructure is typically designed for historical climate conditions (Ch. 12: Transportation, KM 1) and development patterns—for instance, coastal land use—generally do not account for a changing climate (Ch. 5: Land Changes, State of the Sector), resulting in increasing vulnerability to future risks from weather extremes and climate change (Ch. 11: Urban, KM 2). Infrastructure age and deterioration make failure or interrupted service from extreme weather even more likely (Ch. 11: Urban, KM 2). Climate change is expected to increase the costs of maintaining, repairing, and replacing infrastructure, with differences across regions (Ch. 12: Transportation, Regional Summary).

Recent extreme events demonstrate the vulnerabilities of interconnected economic sectors to increasing risks from climate change (see Box 1.3). In 2017, Hurricane Harvey dumped an unprecedented amount of rainfall over the greater Houston area, some of which has been attributed to human-induced climate change (Ch. 2: Climate, Box 2.5). Resulting power outages had cascading effects on critical infrastructure facilities such as hospitals and water and wastewater treatment plants. Reduced oil production and refining capacity in the Gulf of Mexico caused price spikes regionally and nationally from actual and anticipated gasoline shortages (Figure 1.6) (Ch. 17: Complex Systems, KM 1). In the U.S. Caribbean, Hurricanes Irma and Maria caused catastrophic damage to infrastructure, including the complete failure of Puerto Rico's power grid and the loss of power throughout the U.S. Virgin Islands, as well as extensive damage to the region's agricultural industry. The death toll in Puerto Rico grew in the three months following Maria's landfall on the island due in part to the lack of electricity and potable water as well as access to medical facilities and medical care (Ch. 20: U.S. Caribbean, Box 20.1, KM 5).

Climate-related risks to infrastructure, property, and the economy vary across regions. Along the U.S. coastline, public infrastructure and $1 trillion in national wealth held in coastal real estate are threatened by rising sea levels, higher storm surges, and the ongoing increase in high tide flooding (Figures 1.4 and 1.8) (Ch. 8: Coastal, KM 1). Coastal infrastructure provides critical lifelines to the rest of the country, including energy supplies and access to goods and services from overseas trade; increased damage to coastal facilities is expected to result in cascading costs and national impacts (Ch. 8: Coastal, KM 1; Ch. 4: Energy, State of the Sector, KM 1). High tide flooding is projected to become more disruptive and costlier as its frequency, depth, and inland extent grow in the coming decades. Without significant adaptation measures, many coastal cities in the Southeast are expected to experience daily high tide flooding by the end of the century (Ch. 8:

Widespread Impacts from Hurricane Harvey

Figure 1.6: Hurricane Harvey led to widespread flooding and knocked out power to 300,000 customers in Texas in 2017, with cascading effects on critical infrastructure facilities such as hospitals, water and wastewater treatment plants, and refineries. The photo shows Port Arthur, Texas, on August 31, 2017—six days after Hurricane Harvey made landfall along the Gulf Coast. *From Figure 17.2, Ch. 17: Complex Systems (Photo credit: Staff Sgt. Daniel J. Martinez, U.S. Air National Guard).*

Flooding at Fort Calhoun Nuclear Power Plant

Figure 1.7: Floodwaters from the Missouri River surround the Omaha Public Power District's Fort Calhoun Station, a nuclear power plant just north of Omaha, Nebraska, on June 20, 2011. The flooding was the result of runoff from near-record snowfall totals and record-setting rains in late May and early June. A protective berm holding back the floodwaters from the plant failed, which prompted plant operators to transfer offsite power to onsite emergency diesel generators. Cooling for the reactor temporarily shut down, but spent fuel pools were unaffected. *From Figure 22.5, Ch. 22: N. Great Plains (Photo credit: Harry Weddington, U.S. Army Corps of Engineers).*

Norfolk Naval Base at Risk from Rising Seas

Figure 1.8: Low-lying Norfolk, Virginia, houses the world's largest naval base, which supports multiple aircraft carrier groups and is the duty station for thousands of employees. Most of the area around the base lies less than 10 feet above sea level, and local relative sea level is projected to rise between about 2.5 and 11.5 feet by the year 2100 under the Lower and Upper Bound USGCRP sea level rise scenarios, respectively (see Scenario Products section of Appendix 3 for more details on these sea level rise scenarios; see also Ch. 8: Coastal, Case Study "Key Messages in Action—Norfolk, Virginia"). *Photo credit: Mass Communication Specialist 1st Class Christopher B. Stoltz, U.S. Navy.*

Coastal, KM 1; Ch. 19: Southeast, KM 2). Higher sea levels will also cause storm surge from tropical storms to travel farther inland than in the past, impacting more coastal properties and infrastructure (Ch. 8: Coastal: KM 1; Ch. 19: Southeast, KM 2). Oil, natural gas, and electrical infrastructure located along the coasts of the Atlantic Ocean and Gulf of Mexico are at increased risk of damage from rising sea levels and stronger hurricanes; regional disruptions are expected to have national implications (Ch. 4: Energy, State of the Sector, KM 1; Ch.

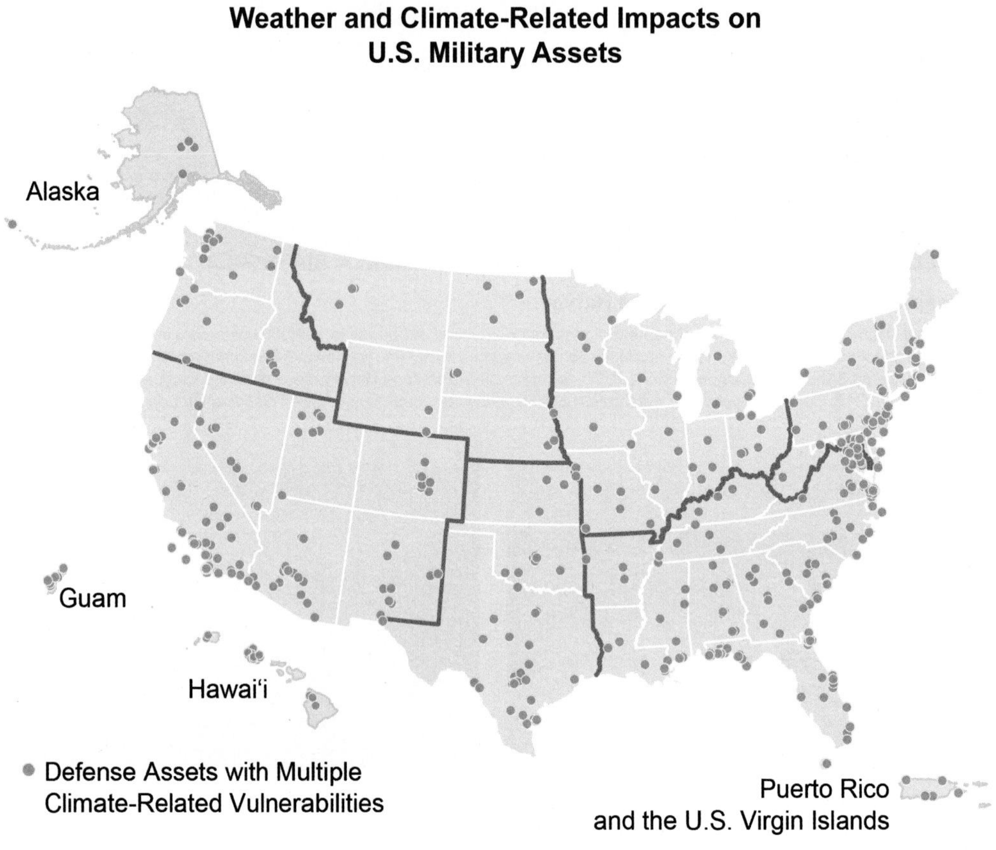

Weather and Climate-Related Impacts on U.S. Military Assets

Alaska

Guam

Hawai'i

● Defense Assets with Multiple Climate-Related Vulnerabilities

Puerto Rico and the U.S. Virgin Islands

Figure 1.9: The Department of Defense (DoD) has significant experience in planning for and managing risk and uncertainty. The effects of climate and extreme weather represent additional risks to incorporate into the Department's various planning and risk management processes. To identify DoD installations with vulnerabilities to climate-related impacts, a preliminary Screening Level Vulnerability Assessment Survey (SLVAS) of DoD sites worldwide was conducted in 2015. The SLVAS responses (shown for the United States; orange dots) yielded a wide range of qualitative information. The highest number of reported effects resulted from drought (782), followed closely by wind (763) and non-storm surge related flooding (706). About 10% of sites indicated being affected by extreme temperatures (351), while flooding due to storm surge (225) and wildfire (210) affected about 6% of the sites reporting. The survey responses provide a preliminary qualitative picture of DoD assets currently affected by severe weather events as well as an indication of assets that may be affected by sea level rise in the future. *Source: adapted from Department of Defense 2018 (http://www.oea.gov/resource/2018-climate-related-risk-dod-infrastructure-initial-vulnerability-assessment-survey-slvas).*

18: Northeast, KM 3; Ch. 19: Southeast, KM 2). Hawai'i and the U.S.-Affiliated Pacific Islands and the U.S. Caribbean also face high risks to critical infrastructure from coastal flooding, erosion, and storm surge (Ch. 4: Energy, State of the Sector; Ch. 20: U.S. Caribbean, KM 3; Ch. 27: Hawai'i & Pacific Islands, KM 3).

In the western United States, increasing wildfire is damaging ranches and rangelands as well as property in cities near the wildland–urban interface. Drier conditions are projected to increase the risk of wildfires and damage to property and infrastructure, including energy production and generation assets and the power grid (Ch. 4: Energy, KM 1; Ch. 11: Urban, Regional Summary; Ch. 24: Northwest, KM 3). In Alaska, thawing of permafrost is responsible for severe damage to roads, buildings, and pipelines that will be costly to replace, especially in remote parts of Alaska. Alaska oil and gas operations are vulnerable to thawing permafrost, sea level rise, and increased coastal exposure due to declining sea ice; however, a longer ice-free season may enhance offshore energy operations and transport (Ch. 4: Energy, State of the Sector; Ch. 26: Alaska, KM 2 and 5). These impacts are expected to grow with continued warming.

U.S. agriculture and the communities it supports are threatened by increases in temperatures, drought, heavy precipitation events, and wildfire on rangelands (Figure 1.10) (Ch. 10: Ag & Rural, KM 1 and 2, Case Study "Groundwater Depletion in the Ogallala Aquifer Region"; Ch. 23: S. Great Plains, KM 1, Case Study "The Edwards Aquifer"). Yields of major U.S. crops (such as corn, soybeans, wheat, rice, sorghum, and cotton) are expected to decline over this century as a consequence of increases in temperatures and possibly changes in water

Conservation Practices Reduce Impact of Heavy Rains

Figure 1.10: Increasing heavy rains are leading to more soil erosion and nutrient loss on midwestern cropland. Integrating strips of native prairie vegetation into row crops has been shown to reduce soil and nutrient loss while improving biodiversity. The inset shows a close-up example of a prairie vegetation strip. *From Figure 21.2, Ch. 21: Midwest (Photo credits: [main photo] Lynn Betts; [inset] Farnaz Kordbacheh).*

availability and disease and pest outbreaks (Ch. 10: Ag & Rural, KM 1). Increases in growing season temperatures in the Midwest are projected to be the largest contributing factor to declines in U.S. agricultural productivity (Ch. 21: Midwest, KM 1). Climate change is also expected to lead to large-scale shifts in the availability and prices of many agricultural products across the world, with corresponding impacts on U.S. agricultural producers and the U.S. economy (Ch. 16: International, KM 1).

Extreme heat poses a significant risk to human health and labor productivity in the agricultural, construction, and other outdoor sectors (Ch. 10: Ag & Rural, KM 3). Under a higher scenario (RCP8.5), almost two billion labor hours are projected to be lost annually by 2090 from the impacts of temperature extremes, costing an estimated $160 billion in lost wages (Ch. 14: Human Health, KM 4). States within the Southeast (Ch. 19: Southeast, KM 4) and Southern Great Plains (Ch. 23: S. Great Plains, KM 4) regions are projected to experience some of the greatest impacts (see Figure 1.21).

Natural Environment and Ecosystem Services

Climate change threatens many benefits that the natural environment provides to society: safe and reliable water supplies, clean air, protection from flooding and erosion, and the use of natural resources for economic, recreational, and subsistence activities. Valued aspects of regional heritage and quality of life tied to the natural environment, wildlife, and outdoor recreation will change with the climate, and as a result, future generations can expect to experience and interact with natural systems in ways that are much different than today. Without significant reductions in greenhouse gas emissions, extinctions and transformative impacts on some ecosystems cannot be avoided, with varying impacts on the economic, recreational, and subsistence activities they support.

Changes affecting the quality, quantity, and availability of water resources, driven in part by climate change, impact people and the environment (Ch. 3: Water, KM 1). Dependable and safe water supplies for U.S. Caribbean, Hawai'i, and U.S.-Affiliated Pacific Island communities and ecosystems are threatened by rising temperatures, sea level rise, saltwater intrusion, and increased risks of drought and flooding (Ch. 3: Water, Regional Summary; Ch. 20: U.S. Caribbean, KM 1; Ch. 27: Hawai'i & Pacific Islands, KM 1). In the Midwest, the occurrence of conditions that contribute to harmful algal blooms, which can result in restrictions to water usage for drinking and recreation, is expected to increase (Ch. 3: Water, Regional Summary; Ch. 21: Midwest, KM 3). In the Southwest, water supplies for people and nature are decreasing during droughts due in part to climate change.

Intensifying droughts, heavier downpours, and reduced snowpack are combining with other stressors such as groundwater depletion to reduce the future reliability of water supplies in the region, with cascading impacts on energy production and other water-dependent sectors (Ch. 3: Water, Regional Summary; Ch. 4: Energy, State of the Sector; Ch. 25: Southwest, KM 5). In the Southern Great Plains, current drought and projected increases in drought length and severity threaten the availability of water for agriculture (Figures 1.11 and 1.12) (Ch. 23: S. Great Plains, KM 1). Reductions in mountain snowpack and shifts in snowmelt timing are expected to reduce hydropower production in the Southwest and the Northwest (Ch. 24: Northwest, KM 3; Ch. 25: Southwest, KM 5). Drought is expected to threaten oil and gas drilling and refining as well as thermoelectric power plants that rely on a steady supply of water for cooling (Ch. 4: Energy, State of the Sector, KM 1; Ch. 22: N. Great Plains, KM 4; Ch. 23: S. Great Plains, KM 2; Ch. 25: Southwest, KM 5).

Impacts of Drought on Texas Agriculture

Figure 1.11: Soybeans in Texas experience the effects of drought in August 2013. During 2010–2015, a multiyear regional drought severely affected agriculture in the Southern Great Plains. One prominent impact was the reduction of irrigation water released for farmers on the Texas coastal plains. *Photo credit: Bob Nichols, USDA.*

Desalination Plants Can Reduce Impacts from Drought in Texas

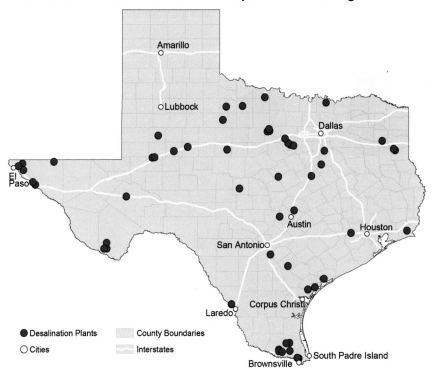

Figure 1.12: Desalination activities in Texas are an important contributor to the state's efforts to meet current and projected water needs for communities, industry, and agriculture. The state's 2017 Water Plan recommended an expansion of desalination to help reduce longer-term risks to water supplies from drought, higher temperatures, and other stressors. There are currently 44 public water supply desalination plants in Texas. *From Figure 23.8, Ch. 23: S. Great Plains (Source: adapted from Texas Water Development Board 2017).*

Tourism, outdoor recreation, and subsistence activities are threatened by reduced snowpack, increases in wildfire activity, and other stressors affecting ecosystems and natural resources (Figures 1.2d, 1.2k, and 1.13) (Ch. 7: Ecosystems, KM 3). Increasing wildfire frequency (Ch. 19: Southeast, Case Study "Prescribed Fire"), pest and disease outbreaks (Ch. 21: Midwest, Case Study "Adaptation in Forestry"), and other stressors are projected to reduce the ability of U.S. forests to support recreation as well as economic and subsistence activities (Ch. 6: Forests, KM 1 and 2; Ch. 19: Southeast, KM 3; Ch. 21: Midwest, KM 2). Increases in wildfire

smoke events driven by climate change are expected to reduce the amount and quality of time spent in outdoor activities (Ch. 13: Air Quality, KM 2; Ch. 24: Northwest, KM 4). Projected declines in snowpack in the western United States and shifts to more precipitation falling as rain than snow in the cold season in many parts of the central and eastern United States are expected to adversely impact the winter recreation industry (Ch. 18: Northeast, KM 1; Ch. 22: N. Great Plains, KM 3; Ch. 24: Northwest, KM 1, Box 24.7). In the Northeast, activities that rely on natural snow and ice cover may not be economically viable by the

Razor Clamming on the Washington Coast

Figure 1.13: Razor clamming draws crowds on the coast of Washington State. This popular recreation activity is expected to decline due to ocean acidification, harmful algal blooms, warmer temperatures, and habitat degradation. *From Figure 24.7, Ch. 24: Northwest (Photo courtesy of Vera Trainer, NOAA).*

end of the century without significant reductions in global greenhouse gas emissions (Ch. 18: Northeast, KM 1). Diminished snowpack, increased wildfire, pervasive drought, flooding, ocean acidification, and sea level rise directly threaten the viability of agriculture, fisheries, and forestry enterprises on tribal lands across the United States and impact tribal tourism and recreation sectors (Ch. 15: Tribes, KM 1).

Climate change has already had observable impacts on biodiversity and ecosystems throughout the United States that are expected to continue. Many species are shifting their ranges (Figure 1.2h), and changes in the timing of important biological events (such as migration and reproduction) are occurring in response to climate change (Ch. 7: Ecosystems, KM 1). Climate change is also aiding the spread of invasive species (Ch. 21: Midwest, Case Study "Adaptation in Forestry"; Ch. 22: N. Great Plains, Case Study "Crow Nation and the Spread of Invasive Species"), recognized as a major driver of biodiversity loss and substantial ecological and economic costs globally (Ch. 7: Ecosystems, Invasive Species). As environmental conditions

change further, mismatches between species and the availability of the resources they need to survive are expected to occur (Ch. 7: Ecosystems, KM 2). Without significant reductions in global greenhouse gas emissions, extinctions and transformative impacts on some ecosystems cannot be avoided in the long term (Ch. 9: Oceans, KM 1). While some new opportunities may emerge from ecosystem changes, economic and recreational opportunities and cultural heritage based around historical use of species or natural resources in many areas are at risk (Ch. 7: Ecosystems, KM 3; Ch. 18: Northeast, KM 1 and 2, Box 18.6).

Ocean warming and acidification pose high and growing risks for many marine organisms, and the impacts of climate change on ocean ecosystems are expected to lead to reductions in important ecosystem services such as aquaculture, fishery productivity, and recreational opportunities (Ch 9: Oceans, KM 2). While climate change impacts on ocean ecosystems are widespread, the scope of ecosystem impacts occurring in tropical and polar areas is greater than anywhere else in the world. Ocean warming is already leading to reductions in vulnerable coral reef and sea ice habitats that support the livelihoods of many communities (Ch. 9: Oceans, KM 1). Decreasing sea ice extent in the Arctic represents a direct loss of important habitat for marine mammals, causing declines in their populations (Figure 1.2f) (Ch. 26: Alaska, Box 26.1). Changes in spring ice melt have affected the ability of coastal communities in Alaska to meet their walrus harvest needs in recent years (Ch. 26: Alaska, KM 1). These changes are expected to continue as sea ice declines further (Ch. 2: Climate, KM 7). In the tropics, ocean warming has already led to widespread coral reef bleaching and/or outbreaks of coral diseases off the coastlines of Puerto

Severe Coral Bleaching Projected for Hawai'i and the U.S.-Affiliated Pacific Islands

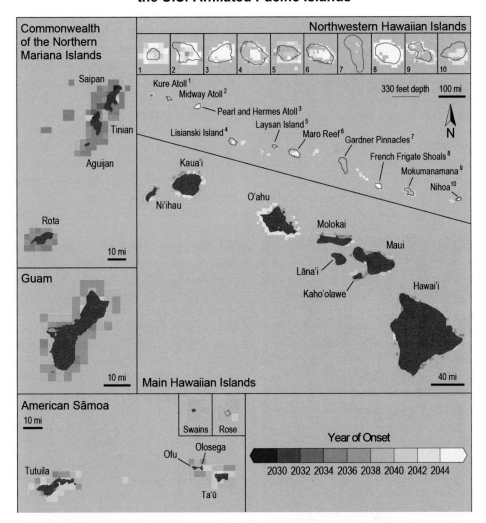

Figure 1.14: The figure shows the years when severe coral bleaching is projected to occur annually in the Hawai'i and U.S.-Affiliated Pacific Islands region under a higher scenario (RCP8.5). Darker colors indicate earlier projected onset of coral bleaching. Under projected warming of approximately 0.5°F per decade, all nearshore coral reefs in the region will experience annual bleaching before 2050. *From Figure 27.10, Ch. 27: Hawai'i & Pacific Islands (Source: NOAA).*

Rico, the U.S. Virgin Islands, Florida, and Hawai'i and the U.S.-Affiliated Pacific Islands (Ch. 20: U.S. Caribbean, KM 2; Ch. 27: Hawai'i & Pacific Islands, KM 4). By mid-century, widespread coral bleaching is projected to occur annually in Hawai'i and the U.S.-Affiliated Pacific Islands (Figure 1.14). Bleaching and ocean acidification are expected to result in loss of reef structure, leading to lower fisheries yields and loss of coastal protection and habitat, with impacts on tourism and livelihoods in both regions (Ch. 20: U.S. Caribbean, KM 2; Ch. 27: Hawai'i & Pacific Islands, KM 4). While some targeted response actions are underway (Figure 1.15), many impacts, including losses of unique coral reef and sea ice ecosystems, can only be avoided by significantly reducing global greenhouse gas emissions, particularly carbon dioxide (Ch. 9: Oceans, KM 1).

Human Health and Well-Being

Higher temperatures, increasing air quality risks, more frequent and intense extreme weather and climate-related events, increases in coastal flooding, disruption of ecosystem services, and other changes increasingly threaten the health and well-being of the American people, particularly populations that are already vulnerable. Future climate change is expected to further disrupt many areas of life, exacerbating existing challenges and revealing new risks to health and prosperity.

Rising temperatures pose a number of threats to human health and quality of life (Figure 1.16). High temperatures in the summer are linked directly to an increased risk of illness and death, particularly among older adults, pregnant women, and children (Ch. 18: Northeast, Box 18.3). With continued warming,

Promoting Coral Reef Recovery

Figure 1.15: Examples of coral farming in the U.S. Caribbean and Florida demonstrate different types of structures used for growing fragments from branching corals. Coral farming is a strategy meant to improve the reef community and ecosystem function, including for fish species. The U.S. Caribbean Islands, Florida, Hawai'i, and the U.S.-Affiliated Pacific Islands face similar threats from coral bleaching and mortality due to warming ocean surface waters and ocean acidification. Degradation of coral reefs is expected to negatively affect fisheries and the economies that depend on them as habitat is lost in both regions. While coral farming may provide some targeted recovery, current knowledge and efforts are not nearly advanced enough to compensate for projected losses from bleaching and acidification. *From Figure 20.11, Ch. 20: U.S. Caribbean (Photo credits: [top left] Carlos Pacheco, U.S. Fish and Wildlife Service; [bottom left] NOAA; [right] Florida Fish and Wildlife).*

Projected Change in Very Hot Days by 2100 in Phoenix, Arizona

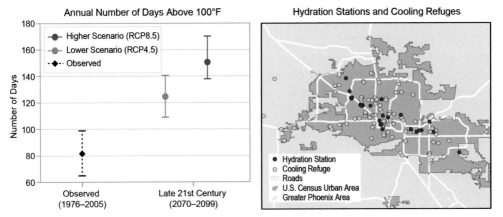

Figure 1.16: (left) The chart shows the average annual number of days above 100°F in Phoenix, Arizona, for 1976–2005, and projections of the average number of days per year above 100°F through the end of the 21st century (2070–2099) under the lower (RCP4.5) and higher (RCP8.5) scenarios. Dashed lines represent the 5th–95th percentile range of annual observed values. Solid lines represent the 5th–95th percentile range of projected model values. (right) The map shows hydration stations and cooling refuges (cooled indoor locations that provide water and refuge from the heat during the day) in Phoenix in August 2017. Such response measures for high heat events are expected to be needed at greater scales in the coming years if the adverse health effects of more frequent and severe heat waves are to be minimized. *Sources: (left) NOAA NCEI, CICS-NC, and LMI; (right) adapted from Southwest Cities Heat Refuges (a project by Arizona State University's Resilient Infrastructure Lab), available at http://www.coolme.today/#phoenix. Data provided by Andrew Fraser and Mikhail Chester, Arizona State University.*

cold-related deaths are projected to decrease and heat-related deaths are projected to increase. In most regions, the increases in heat-related deaths are expected to outpace the reductions in cold-related deaths (Ch. 14: Human Health, KM 1). Rising temperatures are expected to reduce electricity generation capacity while increasing energy demands and costs, which can in turn lead to power outages and blackouts (Ch. 4: Energy, KM 1; Ch. 11: Urban, Regional Summary, Figure 11.2). These changes strain household budgets, increase people's exposure to heat, and limit delivery of medical and social services. Risks from heat stress are higher for people without access to housing with sufficient insulation or air conditioning (Ch. 11: Urban, KM 1).

Changes in temperature and precipitation can increase air quality risks from wildfire and ground-level ozone (smog). Projected increases

in wildfire activity due to climate change would further degrade air quality, resulting in increased health risks and impacts on quality of life (Ch. 13: Air Quality, KM 2; Ch. 14: Human Health, KM 1). Unless counteracting efforts to improve air quality are implemented, climate change is expected to worsen ozone pollution across much of the country, with adverse impacts on human health (Figure 1.21) (Ch. 13: Air Quality, KM 1). Earlier spring arrival, warmer temperatures, changes in precipitation, and higher carbon dioxide concentrations can also increase exposure to airborne pollen allergens. The frequency and severity of allergic illnesses, including asthma and hay fever, are expected to increase as a result of a changing climate (Ch. 13: Air Quality, KM 3).

Rising air and water temperatures and changes in extreme weather and climate-related events are expected to increase exposure to waterborne and foodborne diseases, affecting food and water safety. The geographic range and distribution of disease-carrying insects and pests are projected to shift as climate changes, which could expose more people in North America to ticks that carry Lyme disease and mosquitoes that transmit viruses such as West Nile, chikungunya, dengue, and Zika (Ch. 14: Human Health, KM 1; Ch. 16: International, KM 4).

Mental health consequences can result from exposure to climate- or extreme weather-related events, some of which are projected to intensify as warming continues (Ch. 14: Human Health, KM 1). Coastal city flooding as a result of sea level rise and hurricanes, for example, can result in forced evacuation, with adverse effects on family and community stability as well as mental and physical health (Ch. 11: Urban, KM 1). In urban areas, disruptions in food supply or safety related to extreme weather or climate-related events are expected to disproportionately impact those who already experience food insecurity (Ch. 11: Urban, KM 3).

Indigenous peoples have historical and cultural relationships with ancestral lands, ecosystems, and culturally important species that are threatened by climate change (Ch. 15: Tribes, KM 1; Ch. 19: Southeast, KM 4, Case Study "Mountain Ramps"; Ch. 24: Northwest, KM 5). Climate change is expected to compound existing physical health issues in Indigenous communities, in part due to the loss of traditional foods and practices, and in some cases, the mental stress from permanent community displacement (Ch. 14: Human Health, KM 2; Ch. 15: Tribes, KM 2). Throughout the United States, Indigenous peoples are considering or actively pursuing relocation as an adaptation strategy in response to climate-related disasters, more frequent flooding, loss of land due to erosion, or as livelihoods are compromised by ecosystem shifts linked to climate change (Ch. 15: Tribes, KM 3). In Louisiana, a federal grant is being used to relocate the tribal community of Isle de Jean Charles in response to severe land loss, sea level rise, and coastal flooding (Figure 1.17) (Ch. 19: Southeast, KM 2, Case Study "A Lesson Learned for Community Resettlement"). In Alaska, coastal Native communities are already

Community Relocation—Isle de Jean Charles, Louisiana

Figure 1.17: (left) A federal grant is being used to relocate the tribal community of Isle de Jean Charles, Louisiana, in response to severe land loss, sea level rise, and coastal flooding. *From Figure 15.3, Ch. 15: Tribes (Photo credit: Ronald Stine).* (right) As part of the resettlement of the tribal community of Isle de Jean Charles, residents are working with the Lowlander Center and the State of Louisiana to finalize a plan that reflects the desires of the community. *From Figure 15.4, Ch. 15: Tribes (Photo provided by Louisiana Office of Community Development).*

Adaptation Measures in Kivalina, Alaska

Figure 1.18: A rock revetment was installed in the Alaska Native Village of Kivalina in 2010 to reduce increasing risks from erosion. A new rock revetment wall has a projected lifespan of 15 to 20 years. *From Figure 15.3, Ch. 15: Tribes (Photo credit: ShoreZone. Creative Commons License CC BY 3.0: https://creativecommons.org/licenses/by/3.0/legalcode)*. The inset shows a close-up of the rock wall in 2011. *Photo credit: U.S. Army Corps of Engineers–Alaska District.*

experiencing heightened erosion driven by declining sea ice, rising sea levels, and warmer waters (Figure 1.18). Coastal and river erosion and flooding in some cases will require parts of communities, or even entire communities, to relocate to safer terrain (Ch. 26: Alaska, KM 2). Combined with other stressors, sea level rise, coastal storms, and the deterioration of coral reef and mangrove ecosystems put the long-term habitability of coral atolls in the Hawai'i and U.S.-Affiliated Pacific Islands region at risk, introducing issues of sovereignty, human and national security, and equity (Ch. 27: Hawai'i & Pacific Islands, KM 6).

Reducing the Risks of Climate Change

Climate change is projected to significantly affect human health, the economy, and the environment in the United States, particularly in futures with high greenhouse gas emissions and limited or no adaptation. Recent findings reinforce the fact that without substantial and sustained reductions in greenhouse gas emissions and regional adaptation efforts, there will be substantial and far-reaching changes over the course of the 21st century with negative consequences for a large majority of sectors, particularly towards the end of the century.

The impacts and costs of climate change are already being felt in the United States, and changes in the likelihood or severity of some recent extreme weather events can now be attributed with increasingly higher confidence to human-caused warming (see CSSR, Ch. 3). Impacts associated with human health, such as premature deaths due to extreme temperatures and poor air quality, are some of the most

Box 1.4: How Climate Change Around the World Affects the United States

The impacts of changing weather and climate patterns beyond U.S. international borders affect those living in the United States, often in complex ways that can generate both challenges and opportunities. The International chapter (Ch. 16), new to this edition of the NCA, assesses our current understanding of how global climate change, natural variability, and associated extremes are expected to impact—and in some cases are already impacting—U.S. interests both within and outside of our borders.

Current and projected climate-related impacts on our economy include increased risks to overseas operations of U.S. businesses, disruption of international supply chains, and shifts in the availability and prices of commodities. For example, severe flooding in Thailand in 2011 disrupted the supply chains for U.S. electronics manufacturers (Ch. 16: International, Figure 16.1). U.S. firms are increasingly responding to climate-related risks, including through their financial disclosures and partnerships with environmental groups (Ch. 16: International, KM 1).

Impacts from climate-related events can also undermine U.S. investments in international development by slowing or reversing social and economic progress in developing countries, weakening foreign markets for U.S. exports, and increasing the need for humanitarian assistance and disaster relief efforts. Predictive tools can help vulnerable countries anticipate natural disasters, such as drought, and manage their impacts. For example, the United States and international partners created the Famine Early Warning Systems Network (FEWS NET), which helped avoid severe food shortages in Ethiopia during a historic drought in 2015 (Ch. 16: International, KM 2).

Natural variability and changes in climate increase risks to our national security by affecting factors that can exacerbate conflict and displacement outside of U.S. borders, such as food and water insecurity and commodity price shocks. More directly, our national security is impacted by damage to U.S. military assets such as roads, runways, and waterfront infrastructure from extreme weather and climate-related events (Figures 1.8 and 1.9). The U.S. military is working to both fully understand these threats and incorporate projected climate changes into long-term planning. For example, the Department of Defense has performed a comprehensive scenario-driven examination of climate risks from sea level rise to all of its coastal military sites, including atolls in the Pacific Ocean (Ch. 16: International, KM 3).

Finally, the impacts of climate change are already affecting the ecosystems that span our Nation's borders and the communities that rely on them. International frameworks for the management of our shared resources continue to be restructured to incorporate risks from these impacts. For example, a joint commission that implements water treaties between the United States and Mexico is exploring adaptive water management strategies that account for the effects of climate change and natural variability on Colorado River water (Ch. 16: International, KM 4).

substantial (Ch. 13: Air Quality, KM 1; Ch. 14: Human Health, KM 1 and 4; Ch 29: Mitigation, KM 2). While many sectors face large economic risks from climate change, other impacts can have significant implications for societal or cultural resources. Further, some impacts will very likely be irreversible for thousands of years, including those to species, such as corals (Ch. 9: Oceans, KM 1; Ch. 27: Hawai'i & Pacific Islands, KM 4), or that involve the crossing of thresholds, such as the effects of ice sheet disintegration on accelerated sea level rise, leading to widespread effects on coastal development lasting thousands of years (Ch. 29: Mitigation, KM 2).

Future impacts and risks from climate change are directly tied to decisions made in the present, both in terms of mitigation to reduce emissions of greenhouse gases (or remove carbon dioxide from the atmosphere) and adaptation to reduce risks from today's changed climate conditions and prepare for future impacts. Mitigation and adaptation activities can be considered complementary strategies—mitigation efforts can reduce future risks, while adaptation actions can minimize the consequences of changes that are already happening as a result of past and present greenhouse gas emissions.

Many climate change impacts and economic damages in the United States can be substantially reduced through global-scale reductions in greenhouse gas emissions complemented by regional and local adaptation efforts (Ch 29: Mitigation, KM 4). Our understanding of the magnitude and timing of risks that can be avoided varies by sector, region, and assumptions about how adaptation measures change the exposure and vulnerability of people, livelihoods, ecosystems, and infrastructure. Acting sooner rather than later generally results in lower costs overall for both adaptation and mitigation efforts and can offer other benefits in the near term (Ch. 29: Mitigation, KM 3).

Since the Third National Climate Assessment (NCA3) in 2014, a growing number of states, cities, and businesses have pursued or expanded upon initiatives aimed at reducing greenhouse gas emissions, and the scale of adaptation implementation across the country has increased. However, these efforts do not

yet approach the scale needed to avoid substantial damages to the economy, environment, and human health expected over the coming decades (Ch. 28: Adaptation, KM 1; Ch. 29: Mitigation, KM 1 and 2).

Mitigation

Many activities within the public and private sectors aim for or have the effect of reducing greenhouse gas emissions, such as the increasing use of natural gas in place of coal or the expansion of wind and solar energy to generate electricity. Fossil fuel combustion accounts for approximately 85% of total U.S. greenhouse gas emissions, with agriculture, land-cover change, industrial processes, and methane from fossil fuel extraction and processing as well as from waste (including landfills, wastewater treatment, and composting) accounting for most of the remainder. A number of efforts exist at the federal level to promote low-carbon energy technologies and to increase soil and forest carbon storage.

State, local, and tribal government approaches to mitigating greenhouse gas emissions include comprehensive emissions reduction strategies as well as sector- and technology-specific policies (see Figure 1.19). Since NCA3, private companies have increasingly reported their greenhouse gas emissions, announced emissions reductions targets, implemented actions to achieve those targets, and, in some cases, even put an internal price on carbon. Individuals and other organizations are also making choices every day to reduce their carbon footprints.

Market forces and technological change, particularly within the electric power sector, have contributed to a decline in U.S. greenhouse gas emissions over the past decade. In 2016, U.S.

Mitigation-Related Activities at State and Local Levels

(a)

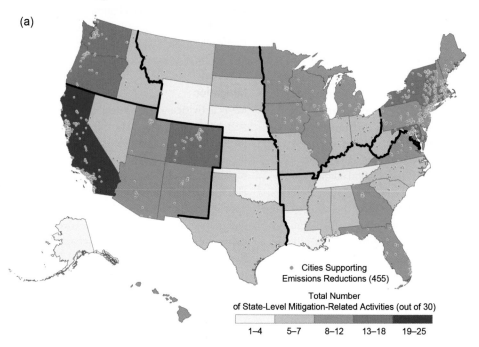

Cities Supporting
Emissions Reductions (455)

Total Number
of State-Level Mitigation-Related Activities (out of 30)

| 1–4 | 5–7 | 8–12 | 13–18 | 19–25 |

(b)

Total State-Level Mitigation-Related Activities by Type

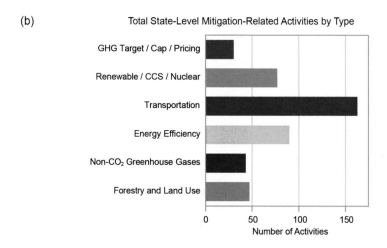

Figure 1.19: (a) The map shows the number of mitigation-related activities at the state level (out of 30 illustrative activities) as well as cities supporting emissions reductions; (b) the chart depicts the type and number of activities by state. Several territories also have a variety of mitigation-related activities, including American Sāmoa, the Federated States of Micronesia, Guam, Northern Mariana Islands, Puerto Rico, and the U.S. Virgin Islands. *From Figure 29.1, Ch. 29: Mitigation (Sources: [a] EPA and ERT, Inc. [b] adapted from America's Pledge 2017).*

emissions were at their lowest levels since 1994. Power sector emissions were 25% below 2005 levels in 2016, the largest emissions reduction for a sector of the American economy over this time. This decline was in large part due to increases in natural gas and renewable energy generation, as well as enhanced energy efficiency standards and programs (Ch. 4: Energy, KM 2). Given these advances in electricity generation, transmission, and distribution, the largest annual sectoral emissions in the United States now come from transportation. As of the writing of this report, business-as-usual (as in, no new policies) projections of U.S. carbon dioxide and other greenhouse gas emissions show flat or declining trajectories over the next decade with a central estimate of about 15% to 20% reduction below 2005 levels by 2025 (Ch. 29: Mitigation, KM 1).

Recent studies suggest that some of the indirect effects of mitigation actions could significantly reduce—or possibly even completely offset—the potential costs associated with cutting greenhouse gas emissions. Beyond reduction of climate pollutants, there are many benefits, often immediate, associated with greenhouse gas emissions reductions, such as improving air quality and public health, reducing crop damages from ozone, and increasing energy independence and security through increased reliance on domestic sources of energy (Ch. 13: Air Quality, KM 4; Ch. 29: Mitigation, KM 4).

Adaptation

Many types of adaptation actions exist, including changes to business operations, hardening infrastructure against extreme weather, and adjustments to natural resource management strategies. Achieving the benefits of adaptation can require upfront investments to achieve longer-term savings, engaging with different stakeholder interests and values, and planning under uncertainty. In many sectors, adaptation can reduce the cost of climate impacts by more than half (Ch. 28: Adaptation, KM 4; Ch. 29: Mitigation, KM 4).

At the time of NCA3's release in 2014, its authors found that risk assessment and planning were underway throughout the United States but that on-the-ground implementation was limited. Since then, the scale and scope of adaptation implementation has increased, including by federal, state, tribal, and local agencies as well as business, academic, and nonprofit organizations (Figure 1.20). While the level of implementation is now higher, it is not yet common nor uniform across the United States, and the scale of implementation for some effects and locations is often considered inadequate to deal with the projected scale of climate change risks. Communities have generally focused on actions that address risks from current climate variability and recent extreme events, such as making buildings and other assets incrementally less sensitive to climate impacts. Fewer communities have focused on actions to address the anticipated scale of future change and emergent threats, such as reducing exposure by preventing building in high-risk locations or retreating from at-risk coastal areas (Ch. 28: Adaptation, KM 1).

Many adaptation initiatives can generate economic and social benefits in excess of their costs in both the near and long term (Ch. 28: Adaptation, KM 4). Damages to infrastructure, such as road and rail networks, are particularly sensitive to adaptation assumptions, with proactive measures that account for future climate risks estimated to be capable of reducing damages by large fractions. More than half of damages to coastal property are estimated to

Five Adaptation Stages and Progress

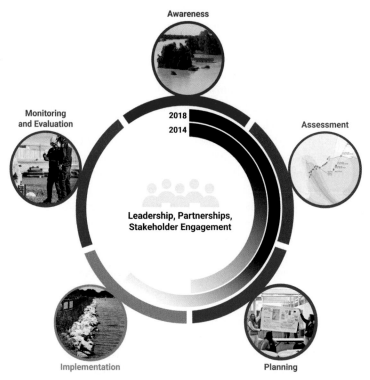

Figure 1.20: Adaptation entails a continuing risk management process. With this approach, individuals and organizations become aware of and assess risks and vulnerabilities from climate and other drivers of change, take actions to reduce those risks, and learn over time. The gray arced lines compare the current status of implementing this process with the status reported by the Third National Climate Assessment in 2014; darker color indicates more activity. *From Figure 28.1, Ch. 28: Adaptation (Source: adapted from National Research Council, 2010. Used with permission from the National Academies Press, © 2010, National Academy of Sciences. Image credits, clockwise from top: National Weather Service; USGS; Armando Rodriguez, Miami-Dade County; Dr. Neil Berg, MARISA; Bill Ingalls, NASA).*

be avoidable through adaptation measures such as shoreline protection and beach replenishment (Ch. 29: Mitigation, KM 4). Considerable guidance is available on actions whose benefits exceed their costs in some sectors (such as adaptation responses to storms and rising seas in coastal zones, to riverine and extreme precipitation flooding, and for agriculture at the farm level), but less so on other actions (such as those aimed at addressing risks to health, biodiversity, and ecosystems services) that may provide significant benefits but are not as well understood (Ch. 28: Adaptation, KM 4).

Effective adaptation can also enhance social welfare in many ways that can be difficult to quantify, including improving economic opportunity, health, equity, national security, education, social connectivity, and sense of place, while safeguarding cultural resources and enhancing environmental quality. Aggregating these benefits into a single monetary value is not always the best approach, and more fundamentally, communities may value benefits differently. Considering various outcomes separately in risk management processes can facilitate participatory planning processes and allow for a specific focus on equity. Prioritizing adaptation actions for populations that face higher risks from climate change, including low-income and marginalized communities, may prove more equitable and lead, for instance, to improved infrastructure in their communities and increased focus on efforts to promote community resilience that can improve their capacity to prepare for, respond to, and recover from disasters (Ch. 28: Adaptation, KM 4).

A significant portion of climate risk can be addressed by integrating climate adaptation into existing investments, policies, and practices. Integration of climate adaptation into decision processes has begun in many areas including financial risk reporting, capital investment planning, engineering standards, military planning, and disaster risk management. A growing number of jurisdictions address climate risk in their land-use, hazard mitigation, capital improvement, and transportation plans, and a small number of cities explicitly link their coastal and hazard mitigation plans using analysis of future climate risks. However, over the course of this century and especially under a higher scenario (RCP8.5), reducing the risks of climate change may require more significant changes to policy and regulations at all scales, community planning, economic and financial systems, technology applications, and ecosystems (Ch. 28: Adaptation, KM 5).

Some sectors are already taking actions that go beyond integrating climate risk into current practices. Faced with substantial climate-induced changes in the future, including new invasive species and shifting ranges for native species, ecosystem managers have already begun to adopt new approaches such as assisted migration and development of wildlife corridors (Ch. 7: Ecosystems, KM 2). Many millions of Americans live in coastal areas threatened by sea level rise; in all but the very lowest sea level rise projections, retreat will become an unavoidable option in some areas along the U.S. coastline (Ch. 8: Coastal, KM 1). The Federal Government has granted funds for the relocation of some communities, including the Biloxi-Chitimacha-Choctaw Tribe from Isle de Jean Charles in Louisiana (Figure 1.17). However, the potential need for millions of people and billions of dollars of coastal infrastructure to be relocated in the future creates challenging legal, financial, and equity issues that have not yet been addressed (Ch. 28: Adaptation, KM 5).

In some areas, lack of historical or current data to inform policy decisions can be a limitation to assessments of vulnerabilities and/or effective adaptation planning. For this National Climate Assessment, this was particularly the case for some aspects of the Alaska, U.S. Caribbean, and Hawai'i and U.S.-Affiliated Pacific Islands regions. In many instances, relying on Indigenous knowledges is among the only current means of reconstructing what has happened in the past. To help communities across the United States learn from one another in their efforts to build resilience to a changing climate, this report highlights common climate-related risks and possible response actions across all regions and sectors.

What Has Happened Since the Last National Climate Assessment?

Our understanding of and experience with climate science, impacts, risks, and adaptation in the United States have grown significantly since the Third National Climate Assessment (NCA3), advancing our knowledge of key processes in the earth system, how human and natural forces are changing them, what the implications are for society, and how we can respond.

Climate Change Impacts in the United States

U.S. National Climate Assessment
U.S. Global Change Research Program

Key Scientific Advances

Detection and Attribution: Significant advances have been made in the attribution of the human influence for individual climate and weather extreme events (see CSSR, Chs. 3, 6, 7, and 8).

Extreme Events and Atmospheric Circulation: How climate change may affect specific types of extreme events in the United States and the extent to which atmospheric circulation in the midlatitudes is changing or is projected to change, possibly in ways not captured by current climate models, are important areas of research where scientific understanding has advanced (see CSSR, Chs. 5, 6, 7, and 9).

Localized Information: As computing resources have grown, projections of future climate from global models are now being conducted at finer scales (with resolution on the order of 15 miles), providing more realistic characterization of intense weather systems, including hurricanes. For the first time in the NCA process, sea level rise projections incorporate geographic variation based on factors such as local land subsidence, ocean currents, and changes in Earth's gravitational field (see CSSR, Chs. 9 and 12).

Ocean and Coastal Waters: Ocean acidification, warming, and oxygen loss are all increasing, and scientific understanding of the severity of their impacts is growing. Both oxygen loss and acidification may be magnified in some U.S. coastal waters relative to the global average, raising the risk of serious ecological and economic consequences (see CSSR, Chs. 2 and 13).

Rapid Changes for Ice on Earth: New observations from many different sources confirm that ice loss across the globe is continuing and, in many cases, accelerating. Since NCA3, Antarctica and Greenland have continued to lose ice mass, with mounting evidence that mass loss is accelerating. Observations continue to show declines in the volume of mountain glaciers around the world. Annual September minimum sea ice extent in the Arctic Ocean has decreased at a rate of 11%–16% per decade since the early 1980s, with accelerating ice loss since 2000. The annual sea ice extent minimum for 2016 was the second lowest on record; the sea ice minimums in 2014 and 2015 were also among the lowest on record (see CSSR, Chs. 1, 11, and 12).

Potential Surprises: Both large-scale shifts in the climate system (sometimes called "tipping points") and compound extremes have the potential to generate outcomes that are difficult to anticipate and may have high consequences. The more the climate changes, the greater the potential for these surprises (see CSSR, Ch. 15).

Extreme Events

Climate change is altering the characteristics of many extreme weather and climate-related events. Some extreme events have already become more frequent, intense, widespread, or of longer duration, and many are expected to continue to increase or worsen, presenting substantial challenges for built, agricultural, and natural systems. Some storm types such as hurricanes, tornadoes, and winter storms are also exhibiting changes that have been linked to climate change, although the current state of the science does not yet permit detailed understanding (see CSSR, Executive Summary). Individual extreme weather and climate-related events—even those that have not been clearly attributed to climate change by scientific analyses—reveal risks to society and vulnerabilities that mirror those we expect in a warmer world. Non-climate stressors (such as land-use changes and shifting demographics) can also amplify the damages associated with extreme events. The National Oceanic and Atmospheric Administration estimates that the United States has experienced 44 billion-dollar weather and climate disasters since 2015 (through April 6, 2018), incurring costs of nearly $400 billion (https://www.ncdc.noaa.gov/billions/).

Hurricanes: The 2017 Atlantic Hurricane season alone is estimated to have caused more than $250 billion in damages and over 250 deaths throughout the U.S. Caribbean, Southeast, and Southern Great Plains. More than 30 inches of rain fell during Hurricane Harvey, affecting 6.9 million people. Hurricane Maria's high winds caused widespread devastation to Puerto Rico's transportation, agriculture, communication, and energy infrastructure. Extreme rainfall of up to 37 inches caused widespread flooding and mudslides across the island. The interruption to commerce and standard living conditions will be sustained for a long period while much of Puerto Rico's infrastructure is rebuilt. Hurricane Irma destroyed 25% of buildings in the Florida Keys.

Damage from Hurricane Maria in San Juan, Puerto Rico

Photo taken during a reconnaissance flight of the island on September 23, 2017. *Photo credit: Sgt. Jose Ahiram Diaz-Ramos, Puerto Rico National Guard.*

Floods: In August 2016, a historic flood resulting from 20 to 30 inches of rainfall over several days devastated a large area of southern Louisiana, causing over $10 billion in damages and 13 deaths. More than 30,000 people were rescued from floodwaters that damaged or destroyed more than 50,000 homes, 100,000 vehicles, and 20,000 businesses. In June 2016, torrential rainfall caused destructive flooding throughout many West Virginia towns, damaging thousands of homes and businesses and causing considerable loss of life. More than 1,500 roads and bridges were damaged or destroyed. The 2015–2016 El Niño poured 11 days of record-setting rainfall on Hawai'i, causing severe urban flooding.

Drought: In 2015, drought conditions caused about $5 billion in damages across the Southwest and Northwest, as well as parts of the Northern Great Plains. California experienced the most severe drought conditions. Hundreds of thousands of acres of farmland remained fallow, and excess groundwater pumping was required to irrigate existing agricultural interests. Two years later, in 2017, extreme drought caused $2.5 billion in agricultural damages across the Northern Great Plains. Field crops, including wheat, were severely damaged, and the lack of feed for cattle forced ranchers to sell off livestock.

Wildfires: During the summer of 2015, over 10.1 million acres—an area larger than the entire state of Maryland—burned across the United States, surpassing 2006 for the highest

The Deadly Carr Fire
The Carr Fire (as seen over Shasta County, California, on August 4, 2018) damaged or destroyed more than 1,500 structures and resulted in several fatalities. *Photo credit: Sgt. Lani O. Pascual, U.S. Army National Guard.*

annual total of U.S. acreage burned since record keeping began in 1960. These wildfire conditions were exacerbated by the preceding drought conditions in several states. The most extensive wildfires occurred in Alaska, where 5 million acres burned within the state. In Montana, wildfires burned in excess of 1 million acres. The costliest wildfires occurred in California, where more than 2,500 structures were destroyed by the Valley and Butte Fires; insured losses alone exceeded $1 billion. In October 2017, a historic firestorm damaged or destroyed more than 15,000 homes, businesses, and other structures across California (see Figure 1.5). The Tubbs, Atlas, Nuns, and Redwood Valley Fires caused a total of 44 deaths, and their combined destruction represents the costliest wildfire event on record.

Tornadoes: In March 2017, a severe tornado outbreak caused damage across much of the Midwest and into the Northeast. Nearly 1 million customers lost power in Michigan alone due to sustained high winds, which affected several states from Illinois to New York.

Heat Waves: Honolulu experienced 24 days of record-setting heat during the 2015–2016 El Niño event. As a result, the local energy utility issued emergency public service announcements to curtail escalating air conditioning use that threatened the electrical grid.

New Aspects of This Report

Hundreds of states, counties, cities, businesses, universities, and other entities are implementing actions that build resilience to climate-related impacts and risks, while also aiming to reduce greenhouse gas emissions. Many of these actions have been informed by new climate-related tools and products developed through the U.S. Global Change Research Program (USGCRP) since NCA3 (see Appendix 3: Scenario Products and Data Tools); we briefly highlight a few of them here. In addition, several structural changes have been introduced to the report and new methods used in response to stakeholder needs for more localized information and to address key gaps identified in NCA3. The Third National Climate Assessment remains a valuable and relevant resource—this report expands upon our knowledge and experience as presented four years ago.

Climate Science Special Report: Early in the development of NCA4, experts and Administration officials recognized that conducting a comprehensive physical science assessment (Volume I) in advance of an impacts assessment (Volume II) would allow one to inform the other. The *Climate Science Special Report,* released in November 2017, is Volume I of NCA4 and represents the most thorough and up-to-date assessment of climate science in the United States and underpins the findings of this report; its findings are summarized in Chapter 2 (Our Changing Climate). See the "Key Scientific Advances" section in this box and Box 2.3 in Chapter 2 for more detail.

Scenario Products: As described in more detail in Appendix 3 (Data Tools & Scenario Products), federal interagency groups developed a suite of high-resolution scenario products that span a range of plausible future changes in key environmental variables through at least 2100. These USGCRP scenario products help ensure consistency across the report and improve the ability to synthesize across chapters. Where possible, authors have used these scenario products to frame uncertainty in future climate as it relates to the risks that are the focus of their chapters. In addition, the Indicators Interagency Working Group has developed an Indicators platform that uses observations or calculations to monitor conditions or trends in the earth system, just as businesses might use the unemployment index as an indicator of economic conditions (see Figure 1.2 and https://www.globalchange.gov/browse/indicators).

Localized Information: With the increased focus on local and regional information in NCA4, USGCRP agencies developed two additional products that not only inform this assessment but can serve as valuable decision-support tools. The first are the State Climate Summaries—a peer-reviewed collection of climate change information covering all ten NCA4 regions at the state level. In addition to standard data on observed and projected climate change, each State Climate Summary contains state-specific changes and their related impacts as well as a suite of complementary graphics (stateclimatesummaries. globalchange.gov). The second product is the U.S. Climate Resilience Toolkit (https:// toolkit.climate.gov/), which offers data-driven tools, information, and subject-matter expertise from across the Federal Government in one easy-to-use location, so Americans are better able to understand the climate-related risks and opportunities impacting their communities and can make more informed decisions on how to respond. In particular, the case studies showcase examples of climate change impacts and accompanying response actions that complement those presented in Figure 1.1 and allow communities to learn how to build resilience from one another.

New Chapters: In response to public feedback on NCA3 and input solicited in the early stages of this assessment, a number of significant structural changes have been made. Most fundamentally, the balance of the report's focus has shifted from national-level chapters to regional chapters in response to a growing desire for more localized information on impacts. Building on this theme, the Great Plains chapter has been split into Northern and Southern chapters (Chapters 22 and 23) along the Kansas–Nebraska border. In addition, the U.S. Caribbean is now featured as a separate region in this report (Chapter 20), focusing on the unique impacts, risks, and response capabilities in Puerto Rico and the U.S. Virgin Islands.

Public input also requested greater international context in the report, which has been addressed through two new additions. A new chapter focuses on topics including the effects of climate change on U.S. trade and businesses, national security, and U.S. humanitarian assistance and disaster relief (Chapter 16). A new international appendix (Appendix 4) presents a number of illustrative examples of how other countries have conducted national climate assessments, putting our own effort into a global context.

Given recent scientific advances, some emerging topics warranted a more visible platform in NCA4. A new chapter on Air Quality (Chapter 13) examines how traditional air pollutants are affected by climate change. A new chapter on Sector Interactions, Multiple Stressors, and Complex Systems (Chapter 17) evaluates climate-related risks to interconnected human and natural systems that are increasingly vulnerable to cascading impacts and highlights advances in analyzing how these systems will interact with and respond to a changing environment (see Box 1.3).

Integrating Economics: This report, to a much greater degree than previous National Climate Assessments, includes broader and more systematic quantification of climate change impacts in economic terms. While this is an emerging body of literature that is not yet reflected in each of the 10 NCA regions, it represents a valuable advancement in our understanding of the financial costs and benefits of climate change impacts. Figure 1.21 provides an illustration of the type of economic information that is integrated throughout this report. It shows the financial damages *avoided* under a lower scenario (RCP4.5) versus a higher scenario (RCP8.5).

New Economic Impact Studies

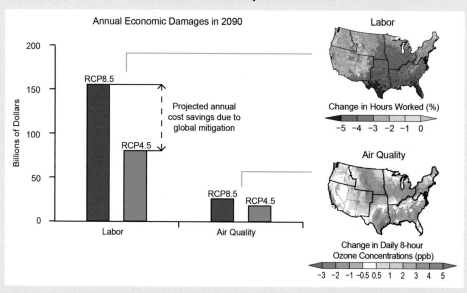

Figure 1.21: Annual economic impact estimates are shown for labor and air quality. The bar graph on the left shows national annual damages in 2090 (in billions of 2015 dollars) for a higher scenario (RCP8.5) and lower scenario (RCP4.5); the difference between the height of the RCP8.5 and RCP4.5 bars for a given category represents an estimate of the economic benefit to the United States from global mitigation action. For these two categories, damage estimates do not consider costs or benefits of new adaptation actions to reduce impacts, and they do not include Alaska, Hawai'i and U.S.-Affiliated Pacific Islands, or the U.S. Caribbean. The maps on the right show regional variation in annual impacts projected under the higher scenario (RCP8.5) in 2090. The map on the top shows the percent change in hours worked in high-risk industries as compared to the period 2003–2007. The hours lost result in economic damages: for example, $28 billion per year in the Southern Great Plains. The map on the bottom is the change in summer-average maximum daily 8-hour ozone concentrations (ppb) at ground-level as compared to the period 1995–2005. These changes in ozone concentrations result in premature deaths: for example, an additional 910 premature deaths each year in the Midwest. *Source: EPA, 2017. Multi-Model Framework for Quantitative Sectoral Impacts Analysis: A Technical Report for the Fourth National Climate Assessment. U.S. Environmental Protection Agency, EPA 430-R-17-001.*

National Topics

Executive Summaries

2 Our Changing Climate

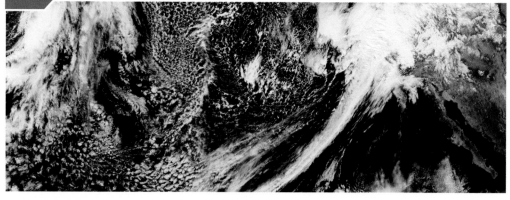

An atmospheric river pours moisture into the western United States in February 2017.

Key Message 1

Observed Changes in Global Climate

Global climate is changing rapidly compared to the pace of natural variations in climate that have occurred throughout Earth's history. Global average temperature has increased by about 1.8°F from 1901 to 2016, and observational evidence does not support any credible natural explanations for this amount of warming; instead, the evidence consistently points to human activities, especially emissions of greenhouse or heat-trapping gases, as the dominant cause.

Key Message 2

Future Changes in Global Climate

Earth's climate will continue to change over this century and beyond. Past mid-century, how much the climate changes will depend primarily on global emissions of greenhouse gases and on the response of Earth's climate system to human-induced warming. With significant reductions in emissions, global temperature increase could be limited to 3.6°F (2°C) or less compared to preindustrial temperatures. Without significant reductions, annual average global temperatures could increase by 9°F (5°C) or more by the end of this century compared to preindustrial temperatures.

Key Message 3

Warming and Acidifying Oceans

The world's oceans have absorbed 93% of the excess heat from human-induced warming since the mid-20th century and are currently absorbing more than a quarter of the carbon dioxide emitted to the atmosphere annually from human activities, making the oceans warmer and more acidic. Increasing sea surface temperatures, rising sea levels, and changing patterns of precipitation, winds, nutrients, and ocean circulation are contributing to overall declining oxygen concentrations in many locations.

Key Message 4

Rising Global Sea Levels

Global average sea level has risen by about 7–8 inches (about 16–21 cm) since 1900, with almost half this rise occurring since 1993 as oceans have warmed and land-based ice has melted. Relative to the year 2000, sea level is very likely to rise 1 to 4 feet (0.3 to 1.3 m) by the end of the century. Emerging science regarding Antarctic ice sheet stability suggests that, for higher scenarios, a rise exceeding 8 feet (2.4 m) by 2100 is physically possible, although the probability of such an extreme outcome cannot currently be assessed.

Key Message 5

Increasing U.S. Temperatures

Annual average temperature over the contiguous United States has increased by 1.2°F (0.7°C) over the last few decades and by 1.8°F (1°C) relative to the beginning of the last century. Additional increases in annual average temperature of about 2.5°F (1.4°C) are expected over the next few decades regardless of future emissions, and increases ranging from 3°F to 12°F (1.6°–6.6°C) are expected by the end of century, depending on whether the world follows a higher or lower future scenario, with proportionally greater changes in high temperature extremes.

Key Message 6

Changing U.S. Precipitation

Annual precipitation since the beginning of the last century has increased across most of the northern and eastern United States and decreased across much of the southern and western United States. Over the coming century, significant increases are projected in winter and spring over the Northern Great Plains, the Upper Midwest, and the Northeast. Observed increases in the frequency and intensity of heavy precipitation events in most parts of the United States are projected to continue. Surface soil moisture over most of the United States is likely to decrease, accompanied by large declines in snowpack in the western United States and shifts to more winter precipitation falling as rain rather than snow.

Key Message 7

Rapid Arctic Change

In the Arctic, annual average temperatures have increased more than twice as fast as the global average, accompanied by thawing permafrost and loss of sea ice and glacier mass. Arctic-wide glacial and sea ice loss is expected to continue; by mid-century, it is very likely that the Arctic will be nearly free of sea ice in late summer. Permafrost is expected to continue to thaw over the coming century as well, and the carbon dioxide and methane released from thawing permafrost has the potential to amplify human-induced warming, possibly significantly.

Key Message 8

Changes in Severe Storms

Human-induced change is affecting atmospheric dynamics and contributing to the poleward expansion of the tropics and the northward shift in Northern Hemisphere winter storm tracks since 1950. Increases in greenhouse gases and decreases in air pollution have contributed to increases in Atlantic hurricane activity since 1970. In the future, Atlantic and eastern North Pacific hurricane rainfall and intensity are projected to increase, as are the frequency and severity of landfalling "atmospheric rivers" on the West Coast.

Key Message 9

Increases in Coastal Flooding

Regional changes in sea level rise and coastal flooding are not evenly distributed across the United States; ocean circulation changes, sinking land, and Antarctic ice melt will result in greater-than-average sea level rise for the Northeast and western Gulf of Mexico under lower scenarios and most of the U.S. coastline other than Alaska under higher scenarios. Since the 1960s, sea level rise has already increased the frequency of high tide flooding by a factor of 5 to 10 for several U.S. coastal communities. The frequency, depth, and extent of tidal flooding are expected to continue to increase in the future, as is the more severe flooding associated with coastal storms, such as hurricanes and nor'easters.

Key Message 10

Long-Term Changes

The climate change resulting from human-caused emissions of carbon dioxide will persist for decades to millennia. Self-reinforcing cycles within the climate system have the potential to accelerate human-induced change and even shift Earth's climate system into new states that are very different from those experienced in the recent past. Future changes outside the range projected by climate models cannot be ruled out, and due to their systematic tendency to underestimate temperature change during past warm periods, models may be more likely to underestimate than to overestimate long-term future change.

For full chapter, including references and Traceable Accounts, see https://nca2018.globalchange.gov/chapter/2/

3 Water

Levee repair along the San Joaquin River in California, February 2017

Key Message 1

Changes in Water Quantity and Quality

Significant changes in water quantity and quality are evident across the country. These changes, which are expected to persist, present an ongoing risk to coupled human and natural systems and related ecosystem services. Variable precipitation and rising temperature are intensifying droughts, increasing heavy downpours, and reducing snowpack. Reduced snow-to-rain ratios are leading to significant differences between the timing of water supply and demand. Groundwater depletion is exacerbating drought risk. Surface water quality is declining as water temperature increases and more frequent high-intensity rainfall events mobilize pollutants such as sediments and nutrients.

Key Message 2

Deteriorating Water Infrastructure at Risk

Deteriorating water infrastructure compounds the climate risk faced by society. Extreme precipitation events are projected to increase in a warming climate and may lead to more severe floods and greater risk of infrastructure failure in some regions. Infrastructure design, operation, financing principles, and regulatory standards typically do not account for a changing climate. Current risk management does not typically consider the impact of compound extremes (co-occurrence of multiple events) and the risk of cascading infrastructure failure.

Key Message 3

Water Management in a Changing Future

Water management strategies designed in view of an evolving future we can only partially anticipate will help prepare the Nation for water- and climate-related risks of the future. Current water management and planning principles typically do not address risk that changes over time, leaving society exposed to more risk than anticipated. While there are examples of promising approaches to manage climate risk, the gap between research and implementation, especially in view of regulatory and institutional constraints, remains a challenge.

Ensuring a reliable supply of clean freshwater to individuals, communities, and ecosystems, together with effective management of floods and droughts, is the foundation of human and ecological health. The water sector is also central to the economy and contributes significantly to the resilience of many other sectors, including agriculture, energy, urban environments, and industry.

Water systems face considerable risk, even without anticipated future climate changes. Limited surface water storage, as well as a limited ability to make use of long-term drought forecasts and to trade water across uses and basins, has led to a significant depletion of aquifers in many regions in the United States. Across the Nation, much of the critical water and wastewater infrastructure is nearing the end of its useful life. To date, no comprehensive assessment exists of the climate-related vulnerability of U.S. water infrastructure (including dams, levees, aqueducts, sewers, and water and wastewater distribution and treatment systems), the potential resulting damages, or the cost of reconstruction and recovery. Paleoclimate information (reconstructions of past climate derived from ice cores or tree rings) shows that over the last 500 years, North America has experienced pronounced wet/dry regime shifts that sometimes persisted for

decades. Because such protracted exposures to extreme floods or droughts in different parts of the country are extraordinary compared to events experienced in the 20th century, they are not yet incorporated in water management principles and practice. Anticipated future climate change will exacerbate this risk in many regions.

A central challenge to water planning and management is learning to plan for plausible future climate conditions that are wider in range than those experienced in the 20th century. Doing so requires approaches that evaluate plans over many possible futures instead of just one, incorporate real-time monitoring and forecast products to better manage extremes when they occur, and update policies and engineering principles with the best available geoscience-based understanding of planetary change. While this represents a break from historical practice, recent examples of adaptation responses undertaken by large water management agencies, including major metropolitan water utilities and the U.S. Army Corps of Engineers, are promising.

For full chapter, including references and Traceable Accounts, see https://nca2018.globalchange.gov/chapter/water.

Depletion of Groundwater in Major U.S. Regional Aquifers

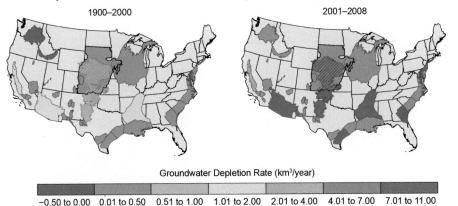

1900–2000   **2001–2008**

Groundwater Depletion Rate (km³/year)

| −0.50 to 0.00 | 0.01 to 0.50 | 0.51 to 1.00 | 1.01 to 2.00 | 2.01 to 4.00 | 4.01 to 7.00 | 7.01 to 11.00 |

(left) Groundwater supplies have been decreasing in the major regional aquifers of the United States over the last century (1900–2000). (right) This decline has accelerated recently (2001–2008) due to persistent droughts in many regions and the lack of adequate surface water storage to meet demands. This decline in groundwater compromises the ability to meet water needs during future droughts and impacts the functioning of groundwater dependent ecosystems (e.g., Kløve et al. 2014). The values shown are net volumetric rates of groundwater depletion (km³ per year) averaged over each aquifer. Subareas of an aquifer may deplete at faster rates or may be actually recovering. Hatching in the figure represents where the High Plains Aquifer overlies the deep, confined Dakota Aquifer. *From Figure 3.2 (Source: adapted from Konikow 2015. Reprinted from Groundwater with permission of the National Groundwater Association. © 2015).*

4 Energy Supply, Delivery, and Demand

Linemen working to restore power in Puerto Rico after Hurricane Maria in 2017

Key Message 1

Nationwide Impacts on Energy

The Nation's energy system is already affected by extreme weather events, and due to climate change, it is projected to be increasingly threatened by more frequent and longer-lasting power outages affecting critical energy infrastructure and creating fuel availability and demand imbalances. The reliability, security, and resilience of the energy system underpin virtually every sector of the U.S. economy. Cascading impacts on other critical sectors could affect economic and national security.

Key Message 2

Changes in Energy System Affect Vulnerabilities

Changes in energy technologies, markets, and policies are affecting the energy system's vulnerabilities to climate change and extreme weather. Some of these changes increase reliability and resilience, while others create additional vulnerabilities. Changes include the following: natural gas is increasingly used as fuel for power plants; renewable resources are becoming increasingly cost competitive with an expanding market share; and a resilient energy supply is increasingly important as telecommunications, transportation, and other critical systems are more interconnected than ever.

Key Message 3

Improving Energy System Resilience

Actions are being taken to enhance energy security, reliability, and resilience with respect to the effects of climate change and extreme weather. This progress occurs through improved data collection, modeling, and analysis to support resilience planning; private and public–private partnerships supporting coordinated action; and both development and deployment of new, innovative energy technologies for adapting energy assets to extreme weather hazards. Although barriers exist, opportunities remain to accelerate the pace, scale, and scope of investments in energy systems resilience.

The Nation's economic security is increasingly dependent on an affordable and reliable supply of energy. Every sector of the economy depends on energy, from manufacturing to agriculture, banking, healthcare, telecommunications, and transportation. Increasingly, climate change and extreme weather events are affecting the energy system, threatening more frequent and longer-lasting power outages and fuel shortages. Such events can have cascading impacts on other critical sectors, potentially affecting the Nation's economic and national security. At the same time, the energy sector is undergoing substantial policy, market, and technology-driven changes that are projected to affect these vulnerabilities.

The impacts of extreme weather and climate change on energy systems will differ across the United States. Low-lying energy facilities and systems located along inland waters or near the coasts are at elevated risk of flooding from more intense precipitation, rising sea levels, and more intense hurricanes. Increases in the severity and frequency of extreme precipitation are projected to affect inland energy infrastructure in every region. Rising temperatures and extreme heat events are projected to reduce the generation capacity of thermoelectric power plants and decrease the efficiency of the transmission grid. Rising temperatures are projected to also drive greater use of air conditioning and increase electricity demand, likely resulting in increases in electricity costs. The increase in annual electricity demand across the country for cooling is offset only marginally by the relatively small decline in electricity demand for heating. Extreme cold events, including ice and snow events, can damage power lines and impact fuel supplies. Severe drought, along with changes in evaporation, reductions in mountain snowpack, and shifting mountain snowmelt timing, is projected to reduce hydropower production and threaten oil and gas drilling and refining, as well as thermoelectric power plants that rely on surface water for cooling. Drier conditions are projected to increase the risk of wildfires and damage to energy production and generation assets and the power grid.

At the same time, the nature of the energy system itself is changing. Low carbon-emitting natural gas generation has displaced coal generation due to the rising production of low-cost, unconventional natural gas, in part supported by federal investment in research and development. In the last 10 years, the share of generation from natural gas increased from 20% to over 30%, while coal has declined from nearly 50% to around 30%. Over this same time, generation from wind and solar has grown from less than 1% to over 5% due to a combination of technological progress, dramatic cost reductions, and federal and state policies.

It is possible to address the challenges of a changing climate and energy system, and both industry and governments at the local, state, regional, federal, and tribal levels are taking actions to improve the resilience of the Nation's energy system. These actions include planning and operational measures that seek to anticipate climate impacts and prevent or respond to damages more effectively, as well as hardening measures to protect assets from damage during extreme events. Resilience actions can have co-benefits, such as developing and deploying new innovative energy technologies that increase resilience and reduce emissions. While steps are being taken, an escalation of the pace, scale, and scope of efforts is needed to ensure the safe and reliable provision of energy and to establish a climate-ready energy system to address present and future risks.

For full chapter, including references and Traceable Accounts, see https://nca2018. globalchange.gov/chapter/energy.

Potential Impacts from Extreme Weather and Climate Change

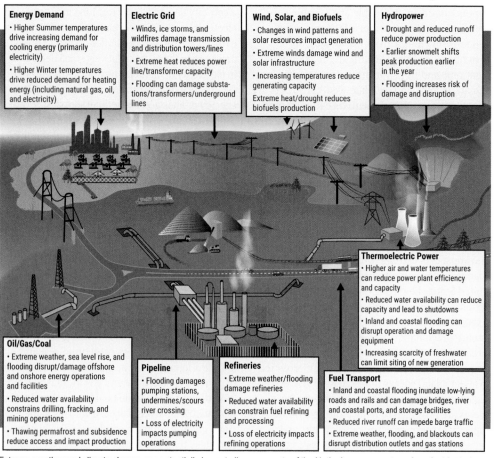

Energy Demand
- Higher Summer temperatures drive increasing demand for cooling energy (primarily electricity)
- Higher Winter temperatures drive reduced demand for heating energy (including natural gas, oil, and electricity)

Electric Grid
- Winds, ice storms, and wildfires damage transmission and distribution towers/lines
- Extreme heat reduces power line/transformer capacity
- Flooding can damage substations/transformers/underground lines

Wind, Solar, and Biofuels
- Changes in wind patterns and solar resources impact generation
- Extreme winds damage wind and solar infrastructure
- Increasing temperatures reduce generating capacity
- Extreme heat/drought reduces biofuels production

Hydropower
- Drought and reduced runoff reduce power production
- Earlier snowmelt shifts peak production earlier in the year
- Flooding increases risk of damage and disruption

Thermoelectric Power
- Higher air and water temperatures can reduce power plant efficiency and capacity
- Reduced water availability can reduce capacity and lead to shutdowns
- Inland and coastal flooding can disrupt operation and damage equipment
- Increasing scarcity of freshwater can limit siting of new generation

Oil/Gas/Coal
- Extreme weather, sea level rise, and flooding disrupt/damage offshore and onshore energy operations and facilities
- Reduced water availability constrains drilling, fracking, and mining operations
- Thawing permafrost and subsidence reduce access and impact production

Pipeline
- Flooding damages pumping stations, undermines/scours river crossing
- Loss of electricity impacts pumping operations

Refineries
- Extreme weather/flooding damage refineries
- Reduced water availability can constrain fuel refining and processing
- Loss of electricity impacts refining operations

Fuel Transport
- Inland and coastal flooding inundate low-lying roads and rails and can damage bridges, river and coastal ports, and storage facilities
- Reduced river runoff can impede barge traffic
- Extreme weather, flooding, and blackouts can disrupt distribution outlets and gas stations

Extreme weather and climate change can potentially impact all components of the Nation's energy system, from fuel (petroleum, coal, and natural gas) production and distribution to electricity generation, transmission, and demand. *From Figure 4.1 (Source: adapted from DOE 2013).*

5 Land Cover and Land-Use Change

Agricultural fields near the Ririe Reservoir in Bonneville, Idaho

Key Message 1

Land-Cover Changes Influence Weather and Climate

Changes in land cover continue to impact local- to global-scale weather and climate by altering the flow of energy, water, and greenhouse gases between the land and the atmosphere. Reforestation can foster localized cooling, while in urban areas, continued warming is expected to exacerbate urban heat island effects.

Key Message 2

Climate Impacts on Land and Ecosystems

Climate change affects land use and ecosystems. Climate change is expected to directly and indirectly impact land use and cover by altering disturbance patterns, species distributions, and the suitability of land for specific uses. The composition of the natural and human landscapes, and how society uses the land, affects the ability of the Nation's ecosystems to provide essential goods and services.

Climate can affect and be affected by changes in land cover (the physical features that cover the land such as trees or pavement) and land use (human management and activities on land, such as mining or recreation). A forest, for instance, would likely include tree cover but could also include areas of recent tree removals currently covered by open grass areas. Land cover and use are inherently coupled: changes in land-use practices can change land cover, and land cover enables specific land uses. Understanding how land cover, use, condition, and management vary in space and time is challenging.

Changes in land cover can occur in response to both human and climate drivers. For example, demand for new settlements often results in the permanent loss of natural and working lands, which can result in localized changes in weather patterns, temperature, and precipitation. Aggregated over large areas, these changes have the potential to influence Earth's climate by altering regional and global circulation patterns, changing the albedo (reflectivity) of Earth's surface, and changing the amount of carbon dioxide (CO_2) in the atmosphere. Conversely, climate change can also influence land cover, resulting in a loss of forest cover from climate-related increases in disturbances, the expansion of woody vegetation into grasslands, and the loss of beaches due to coastal erosion amplified by rises in sea level.

Land use is also changed by both human and climate drivers. Land-use decisions are traditionally based on short-term economic factors. Land-use changes are increasingly being influenced by distant forces due to the globalization of many markets. Land use can also change due to local, state, and national policies, such as programs designed to remove cultivation from highly erodible land to mitigate degradation, legislation to address sea level rise in local comprehensive plans, or policies that reduce the rate of timber harvest on federal lands. Technological innovation has also influenced land-use change, with the expansion of cultivated lands from the development of irrigation technologies and, more recently, decreases in demand for agricultural land due to increases in crop productivity. The recent expansion of oil and gas extraction activities throughout large areas of the United States demonstrates how policy, economics, and technology can collectively influence and change land use and land cover.

Decisions about land use, cover, and management can help determine society's ability to mitigate and adapt to climate change.

For full chapter, including references and Traceable Accounts, see https://nca2018. globalchange.gov/chapter/land-changes.

Changes in Land Cover by Region

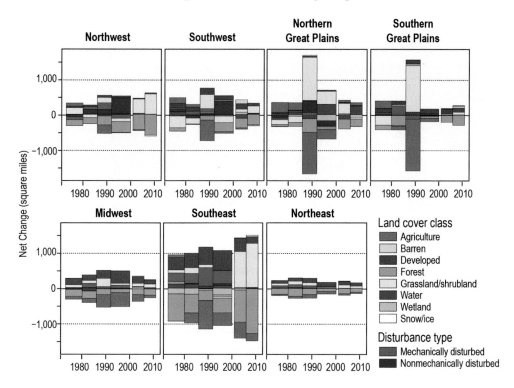

The figure shows the net change in land cover by class in square miles, from 1973 to 2011. Land-cover change has been highly dynamic over space, time, and sector, in response to a range of driving forces. Net change in land cover reveals the trajectory of a class over time. A dramatic example illustrated here is the large decline in agricultural lands in the two Great Plains regions beginning in the mid-1980s, which resulted in large part from the establishment of the Conservation Reserve Program. Over the same period, agriculture also declined in the Southwest region; however, the net decline was largely attributable to prolonged drought conditions, as opposed to changes in federal policy. Data for the period 1973–2000 are from Sleeter et al. (2013) while data from 2001–2011 are from the National Land Cover Database (NLCD). Note: the two disturbance categories used for the 1973–2000 data were not included in the NLCD data for 2001–2011 and largely represent conversions associated with harvest activities (mechanical disturbance) and wildfire (nonmechanical disturbance). Comparable data are unavailable for the U.S. Caribbean, Alaska, and Hawai'i & U.S.-Affiliated Pacific Islands regions, precluding their representation in this figure. *From Figure 5.2 (Source: USGS).*

6 Forests

California's multiyear drought killed millions of trees in low-elevation forests

Key Message 1

Ecological Disturbances and Forest Health

It is very likely that more frequent extreme weather events will increase the frequency and magnitude of severe ecological disturbances, driving rapid (months to years) and often persistent changes in forest structure and function across large landscapes. It is also likely that other changes, resulting from gradual climate change and less severe disturbances, will alter forest productivity and health and the distribution and abundance of species at longer timescales (decades to centuries).

Key Message 2

Ecosystem Services

It is very likely that climate change will decrease the ability of many forest ecosystems to provide important ecosystem services to society. Tree growth and carbon storage are expected to decrease in most locations as a result of higher temperatures, more frequent drought, and increased disturbances. The onset and magnitude of climate change effects on water resources in forest ecosystems will vary but are already occurring in some regions.

Key Message 3

Adaptation

Forest management activities that increase the resilience of U.S. forests to climate change are being implemented, with a broad range of adaptation options for different resources, including applications in planning. The future pace of adaptation will depend on how effectively social, organizational, and economic conditions support implementation.

Forests on public and private lands provide benefits to the natural environment, as well as economic benefits and ecosystem services to people in the United States and globally. The ability of U.S. forests to continue to provide goods and services is threatened by climate change and associated increases in extreme events and disturbances. For example, severe drought and insect outbreaks have killed hundreds of millions of trees across the United States over the past 20 years, and wildfires have burned at least 3.7 million acres annually in all but 3 years from 2000 to 2016. Recent insect-caused mortality appears to be outside the historical context and is likely related to climate change; however, it is unclear if the apparent climate-related increase in fire-caused tree mortality is outside the range of what has been observed over centuries of wildfire occurrence.

A warmer climate will decrease tree growth in most forests that are water limited (for example, low-elevation ponderosa pine forests) but will likely increase growth in forests that are energy limited (for example, subalpine forests, where long-lasting snowpack and cold temperatures limit the growing season). Drought and extreme high temperatures can cause heat-related stress in vegetation and, in turn, reduce forest productivity and increase mortality. The rate of climate warming is likely to influence forest health (that is, the extent to which ecosystem processes are functioning within their range of historic variation) and competition between trees, which will affect the distributions of some species.

Large-scale disturbances (over thousands to hundreds of thousands of acres) that cause rapid change (over days to years) and more gradual climate change effects (over decades) will alter the ability of forests to provide ecosystem services, although alterations will vary greatly depending on the tree species and local biophysical conditions. For example, whereas crown fires (forest fires that spread from treetop to treetop) will cause extensive areas of tree mortality in dense, dry forests in the western United States that have not experienced wildfire for several decades, increased fire frequency is expected to facilitate the persistence of sprouting hardwood species such as quaking aspen in western mountains and fire tolerant pine and hardwood species in the eastern United States (see regional chapters for more detail on variation across the United States). Drought, heavy rainfall, altered snowpack, and changing forest conditions are increasing the frequency of low summer streamflow, winter and spring flooding, and low water quality in some locations, with potential negative impacts on aquatic resources and on water supplies for human communities.

From 1990 to 2015, U.S. forests sequestered 742 teragrams (Tg) of carbon dioxide (CO_2) per year, offsetting approximately 11% of the Nation's CO_2 emissions. U.S. forests are projected to continue to store carbon but at declining rates, as affected by both land use and lower CO_2 uptake as forests get older. However, carbon accumulation in surface soils (at depths of 0–4 inches) can mitigate the declining carbon sink of U.S. forests if reforestation is routinely implemented at large spatial scales.

Implementation of climate-informed resource planning and management on forestlands has progressed significantly over the past decade. The ability of society and resource management to continue to adapt to climate change will be determined primarily by socioeconomic factors and organizational capacity. A viable forest-based workforce can facilitate timely actions that minimize negative effects of climate change. Ensuring the continuing health of forest ecosystems and, where desired and feasible, keeping forestland in forest cover are key challenges for society.

For full chapter, including references and Traceable Accounts, see https://nca2018.globalchange.gov/chapter/forests.

Climate Change Vulnerabilities and Adaptation Options

	Increasing wildfire area burned and fire season length	Increasing drought severity and incidence of insect outbreaks	Lower snowpack, increasing precipitation intensity, and higher winter peakflows	Lower summer streamflows and increasing stream temperatures
Climate Change Vulnerabilities	Increasing wildfire area burned and fire season length	Increasing drought severity and incidence of insect outbreaks	Lower snowpack, increasing precipitation intensity, and higher winter peakflows	Lower summer streamflows and increasing stream temperatures
Adaptation Options	Reduce hazardous fuels with prescribed burning and managed wildfire	Reduce forest stand density to increase tree vigor; plant drought-tolerant species and genotypes	Implement designs for forest road systems that consider increased flooding hazard	Use mapping of projected stream temperatures to set priorities for riparian restoration and coldwater fish conservation

To increase resilience to future stressors and disturbances, examples of adaptation options (risk management) have been developed in response to climate change vulnerabilities in forest ecosystems (risk assessment) in the Pacific Northwest. Vulnerabilities and adaptation options vary among different forest ecosystems. *From Figure 6.7 (Sources: U.S. Forest Service and University of Washington).*

7 Ecosystems, Ecosystem Services, and Biodiversity

Kodiak National Wildlife Refuge, Alaska

Key Message 1

Impacts on Species and Populations

Climate change continues to impact species and populations in significant and observable ways. Terrestrial, freshwater, and marine organisms are responding to climate change by altering individual characteristics, the timing of biological events, and their geographic ranges. Local and global extinctions may occur when climate change outpaces the capacity of species to adapt.

Key Message 2

Impacts on Ecosystems

Climate change is altering ecosystem productivity, exacerbating the spread of invasive species, and changing how species interact with each other and with their environment. These changes are reconfiguring ecosystems in unprecedented ways.

Key Message 3

Ecosystem Services at Risk

The resources and services that people depend on for their livelihoods, sustenance, protection, and well-being are jeopardized by the impacts of climate change on ecosystems. Fundamental changes in agricultural and fisheries production, the supply of clean water, protection from extreme events, and culturally valuable resources are occurring.

Key Message 4

Challenges for Natural Resource Management

Traditional natural resource management strategies are increasingly challenged by the impacts of climate change. Adaptation strategies that are flexible, consider interacting impacts of climate and other stressors, and are coordinated across landscape scales are progressing from theory to application. Significant challenges remain to comprehensively incorporate climate adaptation planning into mainstream natural resource management, as well as to evaluate the effectiveness of implemented actions.

Biodiversity—the variety of life on Earth—provides vital services that support and improve human health and well-being. Ecosystems, which are composed of living things that interact with the physical environment, provide numerous essential benefits to people. These benefits, termed ecosystem services, encompass four primary functions: provisioning materials, such as food and fiber; regulating critical parts of the environment, such as water quality and erosion control; providing cultural services, such as recreational opportunities and aesthetic value; and providing supporting services, such as nutrient cycling. Climate change poses many threats and potential disruptions to ecosystems and biodiversity, as well as to the ecosystem services on which people depend.

Building on the findings of the Third National Climate Assessment (NCA3), this chapter provides additional evidence that climate change is significantly impacting ecosystems and biodiversity in the United States. Mounting evidence also demonstrates that climate change is increasingly compromising the ecosystem services that sustain human communities, economies, and well-being. Both human and natural systems respond to change, but their ability to respond and thrive under new conditions is determined by their adaptive capacity, which may be inadequate to keep pace with rapid change. Our understanding of climate change impacts and the responses of biodiversity and ecosystems has improved since NCA3. The expected consequences of climate change will vary by region, species, and ecosystem type. Management responses are evolving as new tools and approaches are developed and implemented; however, they may not be able to overcome the negative impacts of climate change. Although efforts have been made since NCA3 to incorporate climate adaptation strategies into natural resource management, significant work remains to comprehensively implement climate-informed planning. This chapter presents additional evidence for climate change impacts to biodiversity, ecosystems, and ecosystem services, reflecting increased confidence in the findings reported in NCA3. The chapter also illustrates the complex and interrelated nature of climate change impacts to biodiversity, ecosystems, and the services they provide.

For full chapter, including references and Traceable Accounts, see https://nca2018.globalchange.gov/chapter/ecosystems.

Climate Change, Ecosystems, and Ecosystem Services

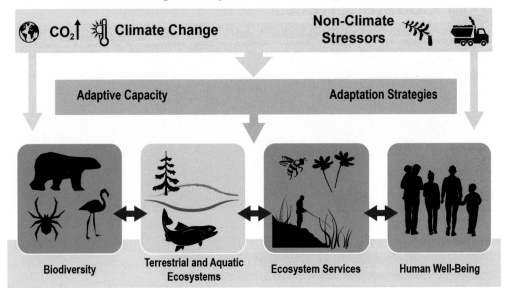

Climate and non-climate stressors interact synergistically on biological diversity, ecosystems, and the services they provide for human well-being. The impact of these stressors can be reduced through the ability of organisms to adapt to changes in their environment, as well as through adaptive management of the resources upon which humans depend. Biodiversity, ecosystems, ecosystem services, and human well-being are interconnected: biodiversity underpins ecosystems, which in turn provide ecosystem services; these services contribute to human well-being. Ecosystem structure and function can also influence the biodiversity in a given area. The use of ecosystem services by humans, and therefore the well-being humans derive from these services, can have feedback effects on ecosystem services, ecosystems, and biodiversity. *From Figure 7.1 (Sources: NOAA, USGS, and DOI).*

8 Coastal Effects

Natural "green barriers" help protect this Florida coastline and infrastructure from severe storms and floods.

Key Message 1

Coastal Economies and Property Are Already at Risk

America's trillion-dollar coastal property market and public infrastructure are threatened by the ongoing increase in the frequency, depth, and extent of tidal flooding due to sea level rise, with cascading impacts to the larger economy. Higher storm surges due to sea level rise and the increased probability of heavy precipitation events exacerbate the risk. Under a higher scenario (RCP8.5), many coastal communities will be transformed by the latter part of this century, and even under lower scenarios (RCP4.5 or RCP2.6), many individuals and communities will suffer financial impacts as chronic high tide flooding leads to higher costs and lower property values. Actions to plan for and adapt to more frequent, widespread, and severe coastal flooding would decrease direct losses and cascading economic impacts.

Key Message 2

Coastal Environments Are Already at Risk

Fisheries, tourism, human health, and public safety depend on healthy coastal ecosystems that are being transformed, degraded, or lost due in part to climate change impacts, particularly sea level rise and higher numbers of extreme weather events. Restoring and conserving coastal ecosystems and adopting natural and nature-based infrastructure solutions can enhance community and ecosystem resilience to climate change, help to ensure their health and vitality, and decrease both direct and indirect impacts of climate change.

Key Message 3

Social Challenges Intensified

As the pace and extent of coastal flooding and erosion accelerate, climate change impacts along our coasts are exacerbating preexisting social inequities, as communities face difficult questions about determining who will pay for current impacts and future adaptation and mitigation strategies and if, how, or when to relocate. In response to actual or projected climate change losses and damages, coastal communities will be among the first in the Nation to test existing climate-relevant legal frameworks and policies against these impacts and, thus, will establish precedents that will affect both coastal and non-coastal regions.

The Coasts chapter of the Third National Climate Assessment, published in 2014, focused on coastal lifelines at risk, economic disruption, uneven social vulnerability, and vulnerable ecosystems. This Coastal Effects chapter of the Fourth National Climate Assessment updates those themes, with a focus on integrating the socioeconomic and environmental impacts and consequences of a changing climate. Specifically, the chapter builds on the threat of rising sea levels exacerbating tidal and storm surge flooding, the state of coastal ecosystems, and the treatment of social vulnerability by introducing the implications for social equity.

U.S. coasts are dynamic environments and economically vibrant places to live and work. As of 2013, coastal shoreline counties were home to 133.2 million people, or 42% of the population. The coasts are economic engines that support jobs in defense, fishing, transportation, and tourism industries; contribute substantially to the U.S. gross domestic product; and serve as hubs of commerce, with seaports connecting the country with global trading partners. Coasts are home to diverse ecosystems such as beaches, intertidal zones, reefs, seagrasses, salt marshes, estuaries, and deltas that support a range of important services including fisheries, recreation, and coastal storm protection. U.S. coasts span three oceans, as well as the Gulf of Mexico, the Great Lakes, and Pacific and Caribbean islands.

The social, economic, and environmental systems along the coasts are being affected by climate change. Threats from sea level rise (SLR) are exacerbated by dynamic processes such as high tide and storm surge flooding (Ch. 19: Southeast, KM 2), erosion (Ch. 26: Alaska, KM 2), waves and their effects, saltwater intrusion into coastal aquifers and elevated groundwater tables (Ch. 27: Hawai'i & Pacific Islands, KM 1; Ch. 3: Water, KM 1), local rainfall (Ch. 3: Water, KM 1), river runoff (Ch. 3: Water, KM 1), increasing water and surface air temperatures (Ch. 9: Oceans, KM 3), and ocean acidification (see Ch. 2: Climate, KM 3 and Ch. 9: Oceans, KM 1, 2, and 3 for more information on ocean acidification, hypoxia, and ocean warming).

Although storms, floods, and erosion have always been hazards, in combination with rising sea levels they now threaten approximately $1 trillion in national wealth held in coastal real estate and the continued viability of coastal communities that depend on coastal water, land, and other resources for economic health and cultural integrity (Ch. 15: Tribes, KM 1 and 2).

For full chapter, including references and Traceable Accounts, see https://nca2018.globalchange.gov/chapter/coastal.

Impacts of the 2017 Hurricane Season
Quintana Perez dumps water from a cooler into floodwaters in the aftermath of Hurricane Irma in Immokalee, Florida. *From Figure 8.6 (Photo credit: AP Photo/Gerald Herbert).*

9 Oceans and Marine Resources

Coral reefs in the U.S. Virgin Islands

Key Message 1

Ocean Ecosystems

The Nation's valuable ocean ecosystems are being disrupted by increasing global temperatures through the loss of iconic and highly valued habitats and changes in species composition and food web structure. Ecosystem disruption will intensify as ocean warming, acidification, deoxygenation, and other aspects of climate change increase. In the absence of significant reductions in carbon emissions, transformative impacts on ocean ecosystems cannot be avoided.

Key Message 2

Marine Fisheries

Marine fisheries and fishing communities are at high risk from climate-driven changes in the distribution, timing, and productivity of fishery-related species. Ocean warming, acidification, and deoxygenation are projected to increase these changes in fishery-related species, reduce catches in some areas, and challenge effective management of marine fisheries and protected species. Fisheries management that incorporates climate knowledge can help reduce impacts, promote resilience, and increase the value of marine resources in the face of changing ocean conditions.

Key Message 3

Extreme Events

Marine ecosystems and the coastal communities that depend on them are at risk of significant impacts from extreme events with combinations of very high temperatures, very low oxygen levels, or very acidified conditions. These unusual events are projected to become more common and more severe in the future, and they expose vulnerabilities that can motivate change, including technological innovations to detect, forecast, and mitigate adverse conditions.

Fourth National Climate Assessment

Americans rely on ocean ecosystems for food, jobs, recreation, energy, and other vital services. Increased atmospheric carbon dioxide levels change ocean conditions through three main factors: warming seas, ocean acidification, and deoxygenation. These factors are transforming ocean ecosystems, and these transformations are already impacting the U.S. economy and coastal communities, cultures, and businesses.

While climate-driven ecosystem changes are pervasive in the ocean, the most apparent impacts are occurring in tropical and polar ecosystems, where ocean warming is causing the loss of two vulnerable habitats: coral reef and sea ice ecosystems. The extent of sea ice in the Arctic is decreasing, which represents a direct loss of important habitat for animals like polar bears and ringed seals that use it for hunting, shelter, migration, and reproduction, causing their abundances to decline (Ch. 26: Alaska, KM 1). Warming has led to mass bleaching and/or outbreaks of coral diseases off the coastlines of Puerto Rico, the U.S. Virgin Islands, Florida, Hawai'i, and the U.S.-Affiliated Pacific Islands (Ch. 20: U.S. Caribbean, KM 2; Ch. 27: Hawai'i & Pacific Islands, KM 4) that threaten reef ecosystems and the people who depend on them. The loss of the recreational benefits alone from coral reefs in the United States is expected to reach $140 billion (discounted at 3% in 2015 dollars) by 2100. Reducing greenhouse gas emissions (for example, under RCP4.5); (see the Scenario Products section of Appendix 3 for more on scenarios) could reduce these cumulative losses by as much as $5.4 billion but will not avoid many ecological and economic impacts.

Ocean warming, acidification, and deoxygenation are leading to changes in productivity, recruitment, survivorship, and, in some cases, active movements of species to track their preferred temperature conditions, with most moving northward or into deeper water with warming oceans. These changes are impacting the distribution and availability of many commercially and recreationally valuable fish and invertebrates. The effects of ocean warming, acidification, and deoxygenation on marine species will interact with fishery management decisions, from seasonal and spatial closures to annual quota setting, allocations, and fish stock rebuilding plans. Accounting for these factors is the cornerstone of climate-ready fishery management. Even without directly accounting for climate effects, precautionary fishery management and better incentives can increase economic benefits and improve resilience.

Short-term changes in weather or ocean circulation can combine with long-term climate trends to produce periods of very unusual ocean conditions that can have significant impacts on coastal communities. Two such events have been particularly well documented: the 2012 marine heat wave in the northwestern Atlantic Ocean and the sequence of warm ocean events between 2014 and 2016 in the northeastern Pacific Ocean, including a large, persistent area of very warm water referred to as "the Blob." Ecosystems within these regions experienced very warm conditions (more than 3.6°F [2°C] above the normal range) that persisted for several months or more. Extreme events in the oceans other than those related to temperature, including ocean acidification and low-oxygen events, can lead to significant disruptions to ecosystems and people, but they can also motivate preparedness and adaptation.

For full chapter, including references and Traceable Accounts, see https://nca2018. globalchange.gov/chapter/oceans.

Extreme Events in U.S. Waters Since 2012

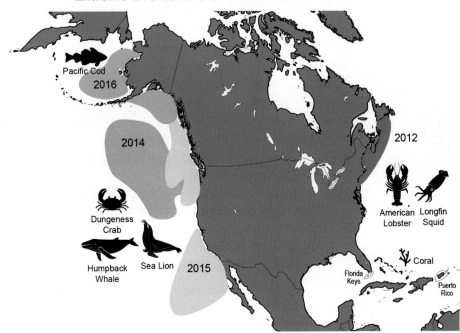

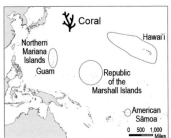

The 2012 North Atlantic heat wave was concentrated in the Gulf of Maine; however, shorter periods with very warm temperatures extended from Cape Hatteras to Iceland during the summer of 2012. American lobster and longfin squid and their associated fisheries were impacted by the event. The North Pacific event began in 2014 and extended toward the shore in 2015 and into the Gulf of Alaska in 2016, leading to a large bloom of toxic algae that impacted the Dungeness crab fishery and contributed directly and indirectly to deaths of sea lions and humpback whales. U.S. coral reefs that experienced moderate to severe bleaching during the 2015–2016 global mass bleaching event are indicated by coral icons. *From Figure 9.3 (Source: Gulf of Maine Research Institute).*

10   Agriculture and Rural Communities

Tyringham, Massachusetts

Key Message 1

Reduced Agricultural Productivity

Food and forage production will decline in regions experiencing increased frequency and duration of drought. Shifting precipitation patterns, when associated with high temperatures, will intensify wildfires that reduce forage on rangelands, accelerate the depletion of water supplies for irrigation, and expand the distribution and incidence of pests and diseases for crops and livestock. Modern breeding approaches and the use of novel genes from crop wild relatives are being employed to develop higher-yielding, stress-tolerant crops.

Key Message 2

Degradation of Soil and Water Resources

The degradation of critical soil and water resources will expand as extreme precipitation events increase across our agricultural landscape. Sustainable crop production is threatened by excessive runoff, leaching, and flooding, which results in soil erosion, degraded water quality in lakes and streams, and damage to rural community infrastructure. Management practices to restore soil structure and the hydrologic function of landscapes are essential for improving resilience to these challenges.

Key Message 3

Health Challenges to Rural Populations and Livestock

Challenges to human and livestock health are growing due to the increased frequency and intensity of high temperature extremes. Extreme heat conditions contribute to heat exhaustion, heatstroke, and heart attacks in humans. Heat stress in livestock results in large economic losses for producers. Expanded health services in rural areas, heat-tolerant livestock, and improved design of confined animal housing are all important advances to minimize these challenges.

Key Message 4

Vulnerability and Adaptive Capacity of Rural Communities

Residents in rural communities often have limited capacity to respond to climate change impacts, due to poverty and limitations in community resources. Communication, transportation, water, and sanitary infrastructure are vulnerable to disruption from climate stressors. Achieving social resilience to these challenges would require increases in local capacity to make adaptive improvements in shared community resources.

In 2015, U.S. agricultural producers contributed $136.7 billion to the economy and accounted for 2.6 million jobs. About half of the revenue comes from livestock production. Other agriculture-related sectors in the food supply chain contributed an additional $855 billion of gross domestic product and accounted for 21 million jobs.

In 2013, about 46 million people, or 15% of the U.S. population, lived in rural counties covering 72% of the Nation's land area. From 2010 to 2015, a historic number of rural counties experienced population declines, and recent demographic trends point to relatively slow employment and population growth in rural areas as well as high rates of poverty. Rural communities, where livelihoods are more tightly interconnected with agriculture, are particularly vulnerable to the agricultural volatility related to climate.

Climate change has the potential to adversely impact agricultural productivity at local, regional, and continental scales through alterations in rainfall patterns, more frequent occurrences of climate extremes (including high temperatures or drought), and altered patterns of pest pressure. Risks associated with climate change depend on the rate and severity of the change and the ability of producers to adapt to changes. These adaptations include altering what is produced, modifying the inputs used for production,

adopting new technologies, and adjusting management strategies.

U.S. agricultural production relies heavily on the Nation's land, water, and other natural resources, and these resources are affected directly by agricultural practices and by climate. Climate change is expected to increase the frequency of extreme precipitation events in many regions in the United States. Because increased precipitation extremes elevate the risk of surface runoff, soil erosion, and the loss of soil carbon, additional protective measures are needed to safeguard the progress that has been made in reducing soil erosion and water quality degradation through the implementation of grassed waterways, cover crops, conservation tillage, and waterway protection strips.

Climate change impacts, such as changes in extreme weather conditions, have a complex influence on human and livestock health. The consequences of climate change on the incidence of drought also impact the frequency and intensity of wildfires, and this holds implications for agriculture and rural communities. Rural populations are the stewards of most of the Nation's forests, watersheds, rangelands, agricultural land, and fisheries. Much of the rural economy is closely tied to the natural environment. Rural residents, and the lands they manage, have the potential to make important economic and conservation contributions to climate change mitigation and adaptation, but their capacity to adapt is impacted by a host of demographic and economic concerns.

For full chapter, including references and Traceable Accounts, see https://nca2018.globalchange. gov/chapter/agriculture-rural.

Agricultural Jobs and Revenue

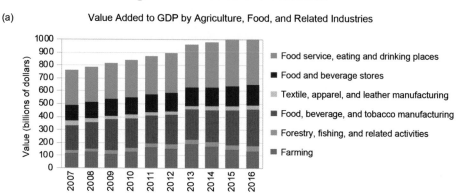

(a) Value Added to GDP by Agriculture, Food, and Related Industries

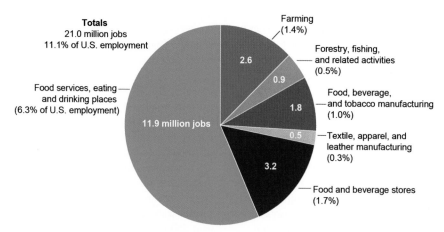

(b) Employment in Agriculture, Food, and Related Industries, 2015

The figure shows (a) the contribution of agriculture and related sectors to the U.S. economy and (b) employment figures in agriculture and related sectors (as of 2015). Agriculture and other food-related value-added sectors account for 21 million full- and part-time jobs and contribute about $1 trillion annually to the United States economy. *From Figure 10.1 (Source: adapted from Kassel et al. 2017).*

Population Changes and Poverty Rates in Rural Counties

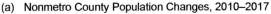

(a) Nonmetro County Population Changes, 2010–2017

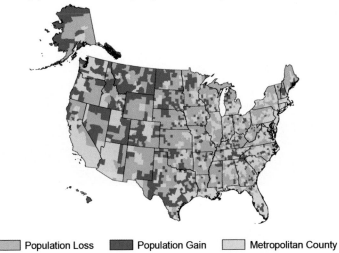

☐ Population Loss   ■ Population Gain   ☐ Metropolitan County

(b) Nonmetro County Poverty Rates, 2011–2015

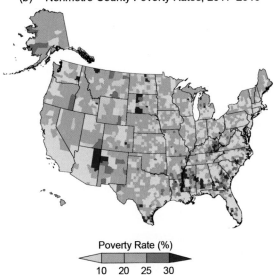

Poverty Rate (%)

10 20 25 30

The figure shows county-level (a) population changes for 2010–2017 and (b) poverty rates for 2011–2015 in rural U.S. communities. Rural populations are migrating to urban regions due to relatively slow employment growth and high rates of poverty. Data for the U.S. Caribbean region were not available at the time of publication of this report. *From Figure 10.2 (Sources: [a] adapted from ERS 2018; [b] redrawn from ERS 2017).*

11 Built Environment, Urban Systems, and Cities

Cleveland, Ohio

Key Message 1

Impacts on Urban Quality of Life

The opportunities and resources in urban areas are critically important to the health and well-being of people who work, live, and visit there. Climate change can exacerbate existing challenges to urban quality of life, including social inequality, aging and deteriorating infrastructure, and stressed ecosystems. Many cities are engaging in creative problem solving to improve quality of life while simultaneously addressing climate change impacts.

Key Message 2

Forward-Looking Design for Urban Infrastructure

Damages from extreme weather events demonstrate current urban infrastructure vulnerabilities. With its long service life, urban infrastructure must be able to endure a future climate that is different from the past. Forward-looking design informs investment in reliable infrastructure that can withstand ongoing and future climate risks.

Key Message 3

Impacts on Urban Goods and Services

Interdependent networks of infrastructure, ecosystems, and social systems provide essential urban goods and services. Damage to such networks from current weather extremes and future climate will adversely affect urban life. Coordinated local, state, and federal efforts can address these interconnected vulnerabilities.

Key Message 4

Urban Response to Climate Change

Cities across the United States are leading efforts to respond to climate change. Urban adaptation and mitigation actions can affect current and projected impacts of climate change and provide near-term benefits. Challenges to implementing these plans remain. Cities can build on local knowledge and risk management approaches, integrate social equity concerns, and join multicity networks to begin to address these challenges.

Urban areas, where the vast majority of Americans live, are engines of economic growth and contain land valued at trillions of dollars. Cities around the United States face a number of challenges to prosperity, such as social inequality, aging and deteriorating infrastructure, and stressed ecosystems. These social, infrastructure, and environmental challenges affect urban exposure and susceptibility to climate change effects.

Urban areas are already experiencing the effects of climate change. Cities differ across regions in the acute and chronic climate stressors they are exposed to and how these stressors interact with local geographic characteristics. Cities are already subject to higher surface temperatures because of the urban heat island effect, which is projected to get stronger. Recent extreme weather events reveal the vulnerability of the built environment (infrastructure such as residential and commercial buildings, transportation, communications, energy, water systems, parks, streets, and landscaping) and its importance to how people live, study, recreate, and work. Heat waves and heavy rainfalls are expected to increase in frequency and intensity. The way city residents respond to such incidents depends on their understanding of risk, their way of life, access to resources, and the communities to which they belong. Infrastructure designed for historical climate trends is vulnerable to future weather extremes and climate change. Investing in forward-looking design can help ensure that infrastructure performs acceptably under changing climate conditions.

Urban areas are linked to local, regional, and global systems. Situations where multiple climate stressors simultaneously affect multiple city sectors, either directly or through system connections, are expected to become more common. When climate stressors affect one sector, cascading effects on other sectors increase risks to residents' health and well-being. Cities across the Nation are taking action in response to climate change. U.S. cities are at the forefront of reducing greenhouse gas emissions and many have begun adaptation planning. These actions build urban resilience to climate change.

For full chapter, including references and Traceable Accounts, see https://nca2018.globalchange.gov/chapter/built-environment.

Projected Change in the Number of Very Hot Days

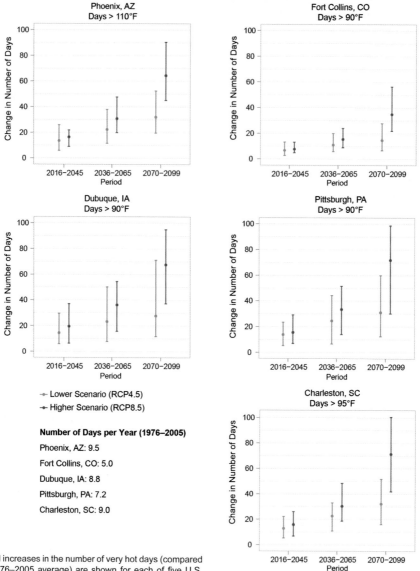

Projected increases in the number of very hot days (compared to the 1976–2005 average) are shown for each of five U.S. cities under lower (RCP4.5) and higher (RCP8.5) scenarios. Here, very hot days are defined as those on which the daily high temperature exceeds a threshold value specific to each of the five U.S. cities shown. Dots represent the modeled median (50th percentile) values, and the vertical bars show the range of values (5th to 95th percentile) from the models used in the analysis. Modeled historical values are shown for the same temperature thresholds, for the period 1976–2005, in the lower left corner of the figure. These and other U.S. cities are projected to see an increase in the number of very hot days over the rest of this century under both scenarios, affecting people, infrastructure, green spaces, and the economy. Increased air conditioning and energy demands raise utility bills and can lead to power outages and blackouts. Hot days can degrade air and water quality, which in turn can harm human health and decrease quality of life. *From Figure 11.2 (Sources: NOAA NCEI, CICS-NC, and LMI).*

12 Transportation

St. Louis, Missouri

Key Message 1

Transportation at Risk

A reliable, safe, and efficient U.S. transportation system is at risk from increases in heavy precipitation, coastal flooding, heat, wildfires, and other extreme events, as well as changes to average temperature. Throughout this century, climate change will continue to pose a risk to U.S. transportation infrastructure, with regional differences.

Key Message 2

Impacts to Urban and Rural Transportation

Extreme events that increasingly impact the transportation network are inducing societal and economic consequences, some of which disproportionately affect vulnerable populations. In the absence of intervention, future changes in climate will lead to increasing transportation challenges, particularly because of system complexity, aging infrastructure, and dependency across sectors.

Key Message 3

Vulnerability Assessments

Engineers, planners, and researchers in the transportation field are showing increasing interest and sophistication in understanding the risks that climate hazards pose to transportation assets and services. Transportation practitioner efforts demonstrate the connection between advanced assessment and the implementation of adaptive measures, though many communities still face challenges and barriers to action.

Transportation is the backbone of economic activity, connecting manufacturers with supply chains, consumers with products and tourism, and people with their workplaces, homes, and communities across both urban and rural landscapes. However, the ability of the transportation sector to perform reliably, safely, and efficiently is undermined by a changing climate. Heavy precipitation, coastal flooding, heat, wildfires, freeze–thaw cycles, and changes in average precipitation and temperature impact individual assets across all modes. These impacts threaten the performance of the entire network, with critical ramifications for economic vitality and mobility, particularly for vulnerable populations and urban infrastructure.

Sea level rise is progressively making coastal roads and bridges more vulnerable and less functional. Many coastal cities across the United States have already experienced an increase in high tide flooding that reduces the functionality of low-elevation roadways, rail, and bridges, often causing costly congestion and damage to infrastructure. Inland transportation infrastructure is highly vulnerable to intense rainfall and flooding. In some regions, the increasing frequency and intensity of heavy precipitation events reduce transportation system efficiency and increase accident risk. High temperatures can stress bridge integrity and have caused more frequent and extended delays to passenger and freight rail systems and air traffic.

Transportation is not only vulnerable to impacts of climate change but also contributes significantly to the causes of climate change. In 2016, the transportation sector became the top contributor to U.S. greenhouse gas emissions. The transportation system is rapidly growing and evolving in response to market demand and innovation. This growth could make climate mitigation and adaptation progressively more challenging to implement and more important to achieve. However, transportation practitioners are increasingly invested in addressing climate risks, as evidenced in more numerous and diverse assessments of transportation sector vulnerabilities across the United States.

For full chapter, including references and Traceable Accounts, see https://nca2018. globalchange.gov/chapter/transportation.

U.S. Transportation Assets and Goals at Risk

Climate Change and Notable Vulnerabilities of Transportation Assets

National Performance Goals at Risk

| Reduced Project Delivery Delays | Safety | Environmental Sustainability | Freight Movement & Economic Vitality | Infrastructure Condition | Congestion Reduction | System Reliability |

Heavy precipitation, coastal flooding, heat, and changes in average precipitation and temperature affect assets (such as roads and bridges) across all modes of transportation. The figure shows major climate-related hazards and the transportation assets impacted. Photos illustrate national performance goals (listed in 23 U.S.C. § 150) that are at risk due to climate-related hazards. *From Figure 12.1 (Source: USGCRP. Photo credits from left to right: JAXPORT, Meredith Fordham Hughes [CC BY-NC 2.0]; Oregon Department of Transportation [CC BY 2.0]; NPS–Mississippi National River and Recreation Area; Flickr user Tom Driggers [CC BY 2.0]; Flickr user Mike Mozart [CC BY 2.0]; Flickr user Jeff Turner [CC BY 2.0]; Flickr user William Garrett [CC BY 2.0] — see https://creativecommons.org/licenses/ for specific Creative Commons licenses).*

13 Air Quality

Carr Fire, Shasta County, California, August 2018

Key Message 1

Increasing Risks from Air Pollution

More than 100 million people in the United States live in communities where air pollution exceeds health-based air quality standards. Unless counteracting efforts to improve air quality are implemented, climate change will worsen existing air pollution levels. This worsened air pollution would increase the incidence of adverse respiratory and cardiovascular health effects, including premature death. Increased air pollution would also have other environmental consequences, including reduced visibility and damage to agricultural crops and forests.

Key Message 2

Increasing Impacts of Wildfires

Wildfire smoke degrades air quality, increasing the health risks to tens of millions of people in the United States. More frequent and severe wildfires due to climate change would further diminish air quality, increase incidences of respiratory illness from exposure to wildfire smoke, impair visibility, and disrupt outdoor recreational activities.

Key Message 3

Increases in Airborne Allergen Exposure

The frequency and severity of allergic illnesses, including asthma and hay fever, are likely to increase as a result of a changing climate. Earlier spring arrival, warmer temperatures, changes in precipitation, and higher carbon dioxide concentrations can increase exposure to airborne pollen allergens.

Key Message 4

Co-Benefits of Greenhouse Gas Mitigation

Many emission sources of greenhouse gases also emit air pollutants that harm human health. Controlling these common emission sources would both mitigate climate change and have immediate benefits for air quality and human health. Because methane is both a greenhouse gas and an ozone precursor, reductions of methane emissions have the potential to simultaneously mitigate climate change and improve air quality.

Unless offset by additional emissions reductions of ozone precursor emissions, there is high confidence that climate change will increase ozone levels over most of the United States, particularly over already polluted areas, thereby worsening the detrimental health and environmental effects due to ozone. The climate penalty results from changes in local weather conditions, including temperature and atmospheric circulation patterns, as well as changes in ozone precursor emissions that are influenced by meteorology. Climate change has already had an influence on ozone concentrations over the United States, offsetting some of the expected ozone benefit from reduced precursor emissions. The magnitude of the climate penalty over the United States could be reduced by mitigating climate change.

Climatic changes, including warmer springs, longer summer dry seasons, and drier soils and vegetation, have already lengthened the wildfire season and increased the frequency of large wildfires. Exposure to wildfire smoke increases the risk of respiratory disease, resulting in adverse impacts to human health. Longer fire seasons and increases in the number of large fires would impair both human health and visibility.

Climate change, specifically rising temperatures and increased carbon dioxide (CO_2) concentrations, can influence plant-based allergens, hay fever, and asthma in three ways: by increasing the duration of the pollen season, by increasing the amount of pollen produced by plants, and by altering the degree of allergic reactions to the pollen.

The energy sector, which includes energy production, conversion, and use, accounts for 84% of greenhouse gas (GHG) emissions in the United States as well as 80% of emissions of nitrogen oxides (NO_x) and 96% of sulfur dioxide, the major precursor of sulfate aerosol. In addition to reducing future warming, reductions in GHG emissions often result in co-benefits (other positive effects, such as improved air quality) and possibly some negative effects (disbenefits) (Ch. 29: Mitigation). Specifically, mitigating GHG emissions can lower emissions of particulate matter (PM), ozone and PM precursors, and other hazardous pollutants, reducing the risks to human health from air pollution.

For full chapter, including references and Traceable Accounts, see https://nca2018.globalchange.gov/chapter/air-quality.

Projected Changes in Summer Season Ozone

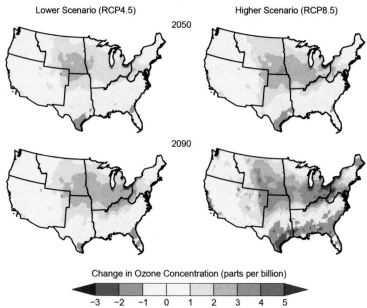

The maps show projected changes in summer averages of the maximum daily 8-hour ozone concentration (as compared to the 1995–2005 average). Summertime ozone is projected to change non-uniformly across the United States based on multiyear simulations from the Community Multiscale Air Quality (CMAQ) modeling system. Those changes are amplified under the higher scenario (RCP8.5) compared with the lower scenario (RCP4.5), as well as at 2090 compared with 2050. Data are not available for Alaska, Hawai'i, U.S.-Affiliated Pacific Islands, and the U.S. Caribbean. *From Figure 13.2 (Source: adapted from EPA 2017).*

14 Human Health

Algal bloom in Lake Erie in the summer of 2015

Key Message 1

Climate Change Affects the Health of All Americans

The health and well-being of Americans are already affected by climate change, with the adverse health consequences projected to worsen with additional climate change. Climate change affects human health by altering exposures to heat waves, floods, droughts, and other extreme events; vector-, food- and waterborne infectious diseases; changes in the quality and safety of air, food, and water; and stresses to mental health and well-being.

Key Message 2

Exposure and Resilience Vary Across Populations and Communities

People and communities are differentially exposed to hazards and disproportionately affected by climate-related health risks. Populations experiencing greater health risks include children, older adults, low-income communities, and some communities of color.

Key Message 3

Adaptation Reduces Risks and Improves Health

Proactive adaptation policies and programs reduce the risks and impacts from climate-sensitive health outcomes and from disruptions in healthcare services. Additional benefits to health arise from explicitly accounting for climate change risks in infrastructure planning and urban design.

Key Message 4

Reducing Greenhouse Gas Emissions Results in Health and Economic Benefits

Reducing greenhouse gas emissions would benefit the health of Americans in the near and long term. By the end of this century, thousands of American lives could be saved and hundreds of billions of dollars in health-related economic benefits gained each year under a pathway of lower greenhouse gas emissions.

Climate-related changes in weather patterns and associated changes in air, water, food, and the environment are affecting the health and well-being of the American people, causing injuries, illnesses, and death. Increasing temperatures, increases in the frequency and intensity of heat waves (since the 1960s), changes in precipitation patterns (especially increases in heavy precipitation), and sea level rise can affect our health through multiple pathways. Changes in weather and climate can degrade air and water quality; affect the geographic range, seasonality, and intensity of transmission of infectious diseases through food, water, and disease-carrying vectors (such as mosquitoes and ticks); and increase stresses that affect mental health and well-being.

Changing weather patterns also interact with demographic and socioeconomic factors, as well as underlying health trends, to influence the extent of the consequences of climate change for individuals and communities. While all Americans are at risk of experiencing adverse climate-related health outcomes, some populations are disproportionately vulnerable.

The risks of climate change for human health are expected to increase in the future, with the extent of the resulting impacts dependent on the effectiveness of adaptation efforts and on the magnitude and pattern of future climate change. Individuals, communities, public health departments, health-related organizations and facilities, and others are taking action to reduce health vulnerability to current climate change and to increase resilience to the risks projected in coming decades.

The health benefits of reducing greenhouse gas emissions could result in economic benefits of hundreds of billions of dollars each year by the end of the century. Annual health impacts and health-related costs are projected to be approximately 50% lower under a lower scenario (RCP4.5) compared to a higher scenario (RCP8.5). These estimates would be even larger if they included the benefits of health outcomes that are difficult to quantify, such as avoided mental health impacts or long-term physical health impacts.

For full chapter, including references and Traceable Accounts, see https://nca2018. globalchange.gov/chapter/health.

Vulnerable Populations

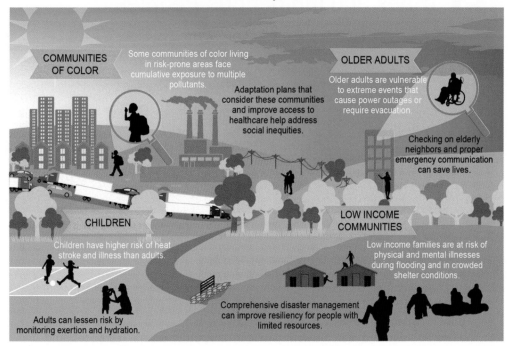

Examples of populations at higher risk of exposure to adverse climate-related health threats are shown along with adaptation measures that can help address disproportionate impacts. When considering the full range of threats from climate change as well as other environmental exposures, these groups are among the most exposed, most sensitive, and have the least individual and community resources to prepare for and respond to health threats. White text indicates the risks faced by those communities, while dark text indicates actions that can be taken to reduce those risks. *From Figure 14.2 (Source: EPA).*

15 Tribes and Indigenous Peoples

Wind River Indian Reservation students collect seeds for a land restoration project.

Key Message 1

Indigenous Livelihoods and Economies at Risk

Climate change threatens Indigenous peoples' livelihoods and economies, including agriculture, hunting and gathering, fishing, forestry, energy, recreation, and tourism enterprises. Indigenous peoples' economies rely on, but face institutional barriers to, their self-determined management of water, land, other natural resources, and infrastructure that will be impacted increasingly by changes in climate.

Key Message 2

Physical, Mental, and Indigenous Values-Based Health at Risk

Indigenous health is based on interconnected social and ecological systems that are being disrupted by a changing climate. As these changes continue, the health of individuals and communities will be uniquely challenged by climate impacts to lands, waters, foods, and other plant and animal species. These impacts threaten sites, practices, and relationships with cultural, spiritual, or ceremonial importance that are foundational to Indigenous peoples' cultural heritages, identities, and physical and mental health.

Key Message 3

Adaptation, Disaster Management, Displacement, and Community-Led Relocations

Many Indigenous peoples have been proactively identifying and addressing climate impacts; however, institutional barriers exist in the United States that severely limit their adaptive capacities. These barriers include limited access to traditional territory and resources and the limitations of existing policies, programs, and funding mechanisms in accounting for the unique conditions of Indigenous communities. Successful adaptation in Indigenous contexts relies on use of Indigenous knowledge, resilient and robust social systems and protocols, a commitment to principles of self-determination, and proactive efforts on the part of federal, state, and local governments to alleviate institutional barriers.

Indigenous peoples in the United States are diverse and distinct political and cultural groups and populations. Though they may be affected by climate change in ways that are similar to others in the United States, Indigenous peoples can also be affected uniquely and disproportionately. Many Indigenous peoples have lived in particular areas for hundreds if not thousands of years. Indigenous peoples' histories and shared experience engender distinct knowledge about climate change impacts and strategies for adaptation. Indigenous peoples' traditional knowledge systems can play a role in advancing understanding of climate change and in developing more comprehensive climate adaptation strategies.

Observed and projected changes of increased wildfire, diminished snowpack, pervasive drought, flooding, ocean acidification, and sea level rise threaten the viability of Indigenous peoples' traditional subsistence and commercial activities that include agriculture, hunting and gathering, fisheries, forestry, energy, recreation, and tourism enterprises. Despite institutional barriers to tribal self-determination stemming from federal trust authority over tribal trust lands, a number of tribes have adaptation plans that include a focus on subsistence and commercial economic activities. Some tribes are also pursuing climate mitigation actions through the development of renewable energy on tribal lands.

Climate impacts to lands, waters, foods, and other plant and animal species threaten cultural heritage sites and practices that sustain intra- and intergenerational relationships built on sharing traditional knowledges, food, and ceremonial or cultural objects. This weakens place-based cultural identities, may worsen historical trauma still experienced by many Indigenous peoples in the United States, and adversely affects mental health and Indigenous values-based understandings of health.

Throughout the United States, climate-related disasters are causing Indigenous communities to consider or actively pursue relocation as an adaptation strategy. Challenges to Indigenous actions to address disaster management and recovery, displacement, and relocation in the face of climate change include economic, social, political, and legal considerations that severely constrain their abilities to respond to rapid ecological shifts and complicate action toward safe and self-determined futures for these communities.

For full chapter, including references and Traceable Accounts, see https://nca2018.globalchange.gov/chapter/tribes.

Indigenous Peoples' Climate Initiatives and Plans

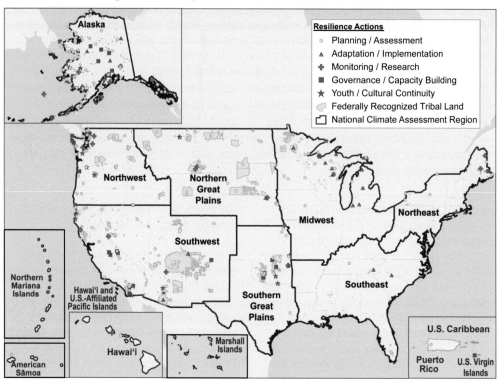

Many Indigenous peoples are taking steps to adapt to climate change impacts. Search the online version of this map by activity type, region, and sector to find more information and links to each project: https://biamaps.doi.gov/nca/. To provide feedback and add new projects for inclusion in the database, see: https://www.bia.gov/bia/ots/tribal-resilience-program/nca/. Thus far, tribal entities in the Northwest have the highest concentration of climate activities (Ch. 24: Northwest). For other case studies of selected tribal adaptation activities, see both the Institute for Tribal Environmental Professionals' Tribal Profiles, and Tribal Case Studies within the U.S. Climate Resilience Toolkit. *From Figure 15.1 (Source: Bureau of Indian Affairs).*

16 Climate Effects on U.S. International Interests

Container ship bringing goods to port

Key Message 1

Economics and Trade

The impacts of climate change, variability, and extreme events outside the United States are affecting and are virtually certain to increasingly affect U.S. trade and economy, including import and export prices and businesses with overseas operations and supply chains.

Key Message 2

International Development and Humanitarian Assistance

The impacts of climate change, variability, and extreme events can slow or reverse social and economic progress in developing countries, thus undermining international aid and investments made by the United States and increasing the need for humanitarian assistance and disaster relief. The United States provides technical and financial support to help developing countries better anticipate and address the impacts of climate change, variability, and extreme events.

Key Message 3

Climate and National Security

Climate change, variability, and extreme events, in conjunction with other factors, can exacerbate conflict, which has implications for U.S. national security. Climate impacts already affect U.S. military infrastructure, and the U.S. military is incorporating climate risks in its planning.

Key Message 4

Transboundary Resources

Shared resources along U.S. land and maritime borders provide direct benefits to Americans and are vulnerable to impacts from a changing climate, variability, and extremes. Multinational frameworks that manage shared resources are increasingly incorporating climate risk in their transboundary decision-making processes.

U.S. international interests, such as economics and trade, international development and humanitarian assistance, national security, and transboundary resources, are affected by impacts from climate change, variability, and extreme events. Long-term changes in climate could lead to large-scale shifts in the global availability and prices of a wide array of agricultural, energy, and other goods, with corresponding impacts on the U.S. economy. Some U.S.-led businesses are already working to reduce their exposure to risks posed by a changing climate.

U.S. investments in international development are sensitive to climate-related impacts and will likely be undermined by more frequent and intense extreme events, such as droughts, floods, and tropical cyclones. These events can impede development efforts and result in greater demand for U.S. humanitarian assistance and disaster relief. In response, the U.S. government has funded adaptation programs that seek to reduce vulnerability to climate impacts in critical sectors.

Climate change, variability, and extreme events increase risks to national security through direct impacts on U.S. military infrastructure and, more broadly, through the relationship between climate-related stress on societies and conflict. Direct linkages between climate and conflict are unclear, but climate variability has been shown to affect conflict through intermediate processes, including resource competition, commodity price shocks, and food insecurity. The U.S. military is working to fully understand these threats and to incorporate projected climate changes into long-term planning.

The impacts of changing weather and climate patterns across U.S. international borders affect those living in the United States. The changes pose new challenges for the management of shared and transboundary resources. Many bilateral agreements and public–private partnerships are incorporating climate risk and adaptive management into their near- and long-term strategies.

U.S. cooperation with international and other national scientific organizations improves access to global information and strategic partnerships, which better positions the Nation to observe, understand, assess, and respond to the impacts associated with climate change, variability, and extremes on national interests both within and outside of U.S. borders.

For full chapter, including references and Traceable Accounts, see https://nca2018. globalchange.gov/chapter/international.

Transboundary Climate-Related Impacts

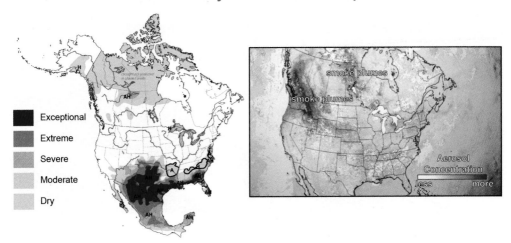

Shown here are examples of climate-related impacts spanning U.S. national borders. (left) The North American Drought Monitor map for June 2011 shows drought conditions along the U.S.–Mexico border. Darker colors indicate greater intensity of drought (the letters A and H indicate agricultural and hydrological drought, respectively). (right) Smoke from Canadian wildfires in 2017 was detected by satellite sensors built to detect aerosols in the atmosphere. The darker orange areas indicate higher concentrations of smoke and hazy conditions moving south from British Columbia to the United States. *From Figure 16.4 (Sources: [left] adapted from NOAA 2018, [right] adapted from NOAA 2018).*

17 Sector Interactions, Multiple Stressors, and Complex Systems

Landslide blocking a road in California

Key Message 1

Interactions Among Sectors

The sectors and systems exposed to climate (for example, energy, water, and agriculture) interact with and depend on one another and other systems less directly exposed to climate (such as the financial sector). In addition, these interacting systems are not only exposed to climate-related stressors such as floods, droughts, and heat waves, they are also subject to a range of non-climate factors, from population movements to economic fluctuations to urban expansion. These interactions can lead to complex behaviors and outcomes that are difficult to predict. It is not possible to fully understand the implications of climate change on the United States without considering the interactions among sectors and their consequences.

Key Message 2

Multisector Risk Assessment

Climate change risk assessment benefits from a multisector perspective, encompassing interactions among sectors and both climate and non-climate stressors. Because such interactions and their consequences can be challenging to identify in advance, effectively assessing multisector risks requires tools and approaches that integrate diverse evidence and that consider a wide range of possible outcomes.

Key Message 3

Management of Interacting Systems

The joint management of interacting systems can enhance the resilience of communities, industries, and ecosystems to climate-related stressors. For example, during drought events, river operations can be managed to balance water demand for drinking water, navigation, and electricity production. Such integrated approaches can help avoid missed opportunities or unanticipated tradeoffs associated with the implementation of management responses to climate-related stressors.

Key Message 4

Advancing Knowledge

Predicting the responses of complex, interdependent systems will depend on developing meaningful models of multiple, diverse systems, including human systems, and methods for characterizing uncertainty.

The world we live in is a web of natural, built, and social systems—from global and regional climate; to the electric grid; to water management systems such as dams, rivers, and canals; to managed and unmanaged forests; and to financial and economic systems. Climate affects many of these systems individually, but they also affect one another, and often in ways that are hard to predict. In addition, while climate-related risks such as heat waves, floods, and droughts have an important influence on these interconnected systems, these systems are also subject to a range of other factors, such as population growth, economic forces, technological change, and deteriorating infrastructure.

A key factor in assessing risk in this context is that it is hard to quantify and predict all the ways in which climate-related stressors might lead to severe or widespread consequences for natural, built, and social systems. A multisector perspective can help identify such critical risks ahead of time, but uncertainties will always remain regarding exactly how consequences will materialize in the future. Therefore, effectively assessing multisector risks requires different tools and approaches than would be applied to understand a single sector by itself.

In interacting systems, management responses within one system influence how other systems respond. Failure to anticipate interdependencies can lead to missed opportunities for managing the risks of climate change; it can also lead to management responses that increase risks to other parts of the system. Despite the challenge of managing system interactions, there are opportunities to learn from experience to guide future risk management decisions.

There is a large gap in the multisector and multiscale tools and frameworks that are available to describe how different human systems interact with one another and with the earth system, and how those interactions affect the total system response to the many stressors they are subject to, including climate-related stressors.

Characterizing the nature of such interactions and building the capacity to model them are important research challenges.

For full chapter, including references and Traceable Accounts, see https://nca2018.globalchange.gov/chapter/complex-systems.

Complex Sectoral Interactions

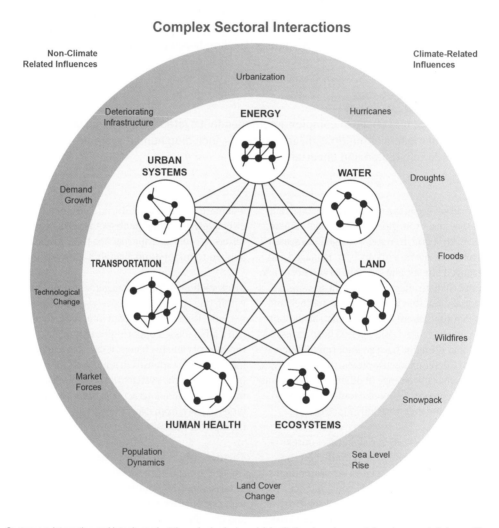

Sectors are interacting and interdependent through physical, social, institutional, environmental, and economic linkages. These sectors and the interactions among them are affected by a range of climate-related and non-climate influences. *From Figure 17.1 (Sources: Pacific Northwest National Laboratory, Arizona State University, and Cornell University).*

Banner Photo Credits

2. Climate: An atmospheric river pours moisture into the western United States in February 2017. *NASA Earth Observatory images by Jesse Allen and Joshua Stevens, using VIIRS data from the Suomi National Polar-orbiting Partnership and IMERG data provided courtesy of the Global Precipitation Mission (GPM) Science Team's Precipitation Processing System (PPS).*

3. Water: Levee repair along the San Joaquin River in California, February 2017. *U.S. Army Corps of Engineers, Sacramento District.*

4. Energy: Linemen working to restore power in Puerto Rico after Hurricane Maria in 2017. *© Jeff Miller/Western Area Power Administration/Flickr. CC BY 2.0,* https://creativecommons.org/licenses/by/2.0/legalcode.

5. Land Changes: Agricultural fields near the Ririe Reservoir in Bonneville, Idaho. *© Sam Beebe/Flickr. CC BY 2.0,* https://creativecommons.org/licenses/by/2.0/legalcode.

6. Forests: California's multiyear drought killed millions of trees in low-elevation forests. *Nathan Stephenson/U.S. Geological Survey.*

7. Ecosystems: Kodiak National Wildlife Refuge, Alaska. *Lisa Hupp/U.S. Fish and Wildlife Service.*

8. Coastal: Natural "green barriers" help protect this Florida coastline and infrastructure from severe storms and floods. *NOAA.*

9. Oceans: Coral reefs in the U.S. Virgin Islands. *NOAA Coral Reef Conservation Program.*

10. Ag & Rural: Tyringham, Massachusetts. *© DenisTangneyJr/E+/Getty Images.*

11. Urban: Cleveland, Ohio. *© Erik Drost/Flickr. CC BY 2.0,* https://creativecommons.org/licenses/by/2.0/legalcode.

12. Transportation: St. Louis, Missouri. *© Cathy Morrison/Missouri Department of Transportation. CC BY-NC-SA 2.0,* https://creativecommons.org/licenses/by-nc-sa/2.0/legalcode.

13. Air Quality: Carr Fire, Shasta County, California, August 2018. *Sgt. Lani O. Pascual/U.S. Army National Guard.*

14. Human Health: Algal bloom in Lake Erie in summer 2015. *NOAA Great Lakes Environmental Research Laboratory.*

15. Tribes: Wind River Indian Reservation students collect seeds for a land restoration project. *U.S. Department of the Interior/Bureau of Land Management Wyoming.*

16: International: Container ship bringing goods to port. *© wissanu01/iStock/Getty Images.*

17. Complex Systems: Landslide blocking a road in California. *© gece33/E+/Getty Images.*

Note: Photos have been cropped from their original size in order to fit the report template.

Regions

18 Northeast

Bartram Bridge in Pennsylvania

Key Message 1

Changing Seasons Affect Rural Ecosystems, Environments, and Economies

The seasonality of the Northeast is central to the region's sense of place and is an important driver of rural economies. Less distinct seasons with milder winter and earlier spring conditions are already altering ecosystems and environments in ways that adversely impact tourism, farming, and forestry. The region's rural industries and livelihoods are at risk from further changes to forests, wildlife, snowpack, and streamflow.

Key Message 2

Changing Coastal and Ocean Habitats, Ecosystem Services, and Livelihoods

The Northeast's coast and ocean support commerce, tourism, and recreation that are important to the region's economy and way of life. Warmer ocean temperatures, sea level rise, and ocean acidification threaten these services. The adaptive capacity of marine ecosystems and coastal communities will influence ecological and socioeconomic outcomes as climate risks increase.

Key Message 3

Maintaining Urban Areas and Communities and Their Interconnectedness

The Northeast's urban centers and their interconnections are regional and national hubs for cultural and economic activity. Major negative impacts on critical infrastructure, urban economies, and nationally significant historic sites are already occurring and will become more common with a changing climate.

Key Message 4

Threats to Human Health

Changing climate threatens the health and well-being of people in the Northeast through more extreme weather, warmer temperatures, degradation of air and water quality, and sea level rise. These environmental changes are expected to lead to health-related impacts and costs, including additional deaths, emergency room visits and hospitalizations, and a lower quality of life. Health impacts are expected to vary by location, age, current health, and other characteristics of individuals and communities.

Key Message 5

Adaptation to Climate Change Is Underway

Communities in the Northeast are proactively planning and implementing actions to reduce risks posed by climate change. Using decision support tools to develop and apply adaptation strategies informs both the value of adopting solutions and the remaining challenges. Experience since the last assessment provides a foundation to advance future adaptation efforts.

The distinct seasonality of the Northeast's climate supports a diverse natural landscape adapted to the extremes of cold, snowy winters and warm to hot, humid summers. This natural landscape provides the economic and cultural foundation for many rural communities, which are largely supported by a diverse range of agricultural, tourism, and natural resource-dependent industries (see Ch. 10: Ag & Rural, Key Message 4). The recent dominant trend in precipitation throughout the Northeast has been towards increases in rainfall intensity, with increases in intensity exceeding those in other regions of the contiguous United States. Further increases in rainfall intensity are expected, with increases in total precipitation expected during the winter and spring but with little change in the summer. Monthly precipitation in the Northeast is projected to be about 1 inch greater for December through April by end of century (2070–2100) under the higher scenario (RCP8.5).

Ocean and coastal ecosystems are being affected by large changes in a variety of climate-related environmental conditions. These ecosystems support fishing and aquaculture, tourism and recreation, and coastal communities. Observed and projected increases in temperature, acidification, storm frequency and intensity, and sea levels are of particular concern for coastal and ocean ecosystems, as well as local communities and their interconnected social and economic systems. Increasing temperatures and changing seasonality on the Northeast Continental Shelf have affected marine organisms and the ecosystem in various ways. The warming trend experienced in the Northeast Continental Shelf has been associated with many fish and invertebrate species moving northward and to greater depths. Because of the diversity of the Northeast's coastal landscape, the impacts from storms and sea level rise will vary at different locations along the coast.

Northeastern cities, with their abundance of concrete and asphalt and relative lack of vegetation, tend to have higher temperatures than surrounding regions due to the urban heat island effect. During extreme heat events, nighttime temperatures in the region's big cities are generally several degrees higher than surrounding regions, leading to higher risk of heat-related death. Urban areas are at risk for large numbers of evacuated and displaced populations and damaged infrastructure due to both extreme precipitation events and recurrent flooding, potentially requiring significant emergency response efforts and consideration of a long-term commitment to rebuilding and adaptation, and/or support for relocation where needed. Much of the infrastructure in the Northeast, including drainage and sewer systems, flood and storm protection assets, transportation systems, and power supply, is nearing the end of its planned life expectancy. Climate-related disruptions will only exacerbate existing issues with aging infrastructure. Sea level rise has amplified storm impacts in the Northeast (Key Message 2), contributing to higher surges that extend farther inland, as demonstrated in New York City in the aftermath of Superstorm Sandy in 2012. Service and resource supply infrastructure in the Northeast is at increasing risk of disruption, resulting in lower quality of life, economic declines, and increased social inequality. Loss of public services affects the capacity of communities to function as administrative and economic centers and triggers disruptions of interconnected supply chains (Ch. 16: International, Key Message 1).

Increases in annual average temperatures across the Northeast range from less than 1°F (0.6°C) in West Virginia to about 3°F (1.7°C) or more in New England since 1901. Although the relative risk of death on very hot days is lower today than it was a few decades ago, heat-related illness and death remain significant public health problems in the Northeast. For example, a study in New York City estimated that in 2013 there were 133 excess deaths due to extreme heat. These projected increases in temperature are expected to lead to substantially more premature deaths, hospital admissions, and emergency department visits across the Northeast. For example, in the Northeast we can expect approximately 650 additional premature deaths per year from extreme heat by the year 2050 under either a lower (RCP4.5) or higher (RCP8.5) scenario and from 960 (under RCP4.5) to 2,300 (under RCP8.5) more premature deaths per year by 2090.

Communities, towns, cities, counties, states, and tribes across the Northeast are engaged in efforts to build resilience to environmental challenges and adapt to a changing climate. Developing and implementing climate adaptation strategies in daily practice often occur in collaboration with state and federal agencies. Advances in rural towns, cities, and suburban areas include low-cost adjustments of existing building codes and standards. In coastal areas, partnerships among local communities and federal and state agencies leverage federal adaptation tools and decision support frameworks. Increasingly, cities and towns across the Northeast are developing or implementing plans for adaptation and resilience in the face of changing climate. The approaches are designed to maintain and enhance the everyday lives of residents and promote economic development. In some cities, adaptation planning has been used to respond to present and future challenges in the built environment. Regional efforts have recommended changes in design standards when building, replacing, or retrofitting infrastructure to account for a changing climate.

For full chapter, including references and Traceable Accounts, see https://nca2018. globalchange.gov/chapter/northeast.

Lengthening of the Freeze-Free Period

| Last Spring Freeze | First Fall Freeze |

2040–2069, Lower Scenario (RCP4.5)

2040–2069, Higher Scenario (RCP8.5)

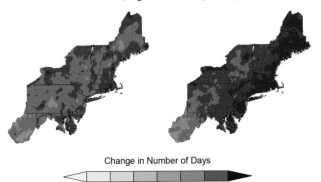

2070–2099, Higher Scenario (RCP8.5)

Change in Number of Days

6   10   14   18   22   26   30

These maps show projected shifts in the date of the last spring freeze (left column) and the date of the first fall freeze (right column) for the middle of the century (as compared to 1979–2008) under the lower scenario (RCP4.5; top row) and the higher scenario (RCP8.5; middle row). The bottom row shows the shift in these dates for the end of the century under the higher scenario. By the middle of the century, the freeze-free period across much of the Northeast is expected to lengthen by as much as two weeks under the lower scenario and by two to three weeks under the higher scenario. By the end of the century, the freeze-free period is expected to increase by at least three weeks over most of the region. *From Figure 18.3 (Source: adapted from Wolfe et al. 2018).*

Coastal Impacts of Climate Change

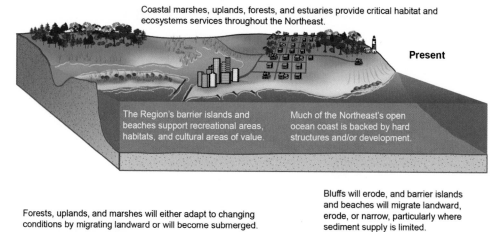

Coastal marshes, uplands, forests, and estuaries provide critical habitat and ecosystems services throughout the Northeast.

Present

The Region's barrier islands and beaches support recreational areas, habitats, and cultural areas of value.

Much of the Northeast's open ocean coast is backed by hard structures and/or development.

Forests, uplands, and marshes will either adapt to changing conditions by migrating landward or will become submerged.

Bluffs will erode, and barrier islands and beaches will migrate landward, erode, or narrow, particularly where sediment supply is limited.

Possible Future

Barrier islands are likely to erode and narrow, especially where sediment supply is limited.

Coastal erosion and flooding will require ongoing efforts to protect or adapt existing development.

(top) The northeastern coastal landscape is composed of uplands and forested areas, wetlands and estuarine systems, mainland and barrier beaches, bluffs, headlands, and rocky shores, as well as developed areas, all of which provide a variety of important services to people and species. (bottom) Future impacts from intense storm activity and sea level rise will vary across the landscape, requiring a variety of adaptation strategies if people, habitats, traditions, and livelihoods are to be protected. *From Figure 18.7 (Source: U.S. Geological Survey).*

19 Southeast

Red mangrove in Titusville, Florida

Key Message 1

Urban Infrastructure and Health Risks

Many southeastern cities are particularly vulnerable to climate change compared to cities in other regions, with expected impacts to infrastructure and human health. The vibrancy and viability of these metropolitan areas, including the people and critical regional resources located in them, are increasingly at risk due to heat, flooding, and vector-borne disease brought about by a changing climate. Many of these urban areas are rapidly growing and offer opportunities to adopt effective adaptation efforts to prevent future negative impacts of climate change.

Key Message 2

Increasing Flood Risks in Coastal and Low-Lying Regions

The Southeast's coastal plain and inland low-lying regions support a rapidly growing population, a tourism economy, critical industries, and important cultural resources that are highly vulnerable to climate change impacts. The combined effects of changing extreme rainfall events and sea level rise are already increasing flood frequencies, which impacts property values and infrastructure viability, particularly in coastal cities. Without significant adaptation measures, these regions are projected to experience daily high tide flooding by the end of the century.

Fourth National Climate Assessment

Key Message 3

Natural Ecosystems Will Be Transformed

The Southeast's diverse natural systems, which provide many benefits to society, will be transformed by climate change. Changing winter temperature extremes, wildfire patterns, sea levels, hurricanes, floods, droughts, and warming ocean temperatures are expected to redistribute species and greatly modify ecosystems. As a result, the ecological resources that people depend on for livelihood, protection, and well-being are increasingly at risk, and future generations can expect to experience and interact with natural systems that are much different than those that we see today.

Key Message 4

Economic and Health Risks for Rural Communities

Rural communities are integral to the Southeast's cultural heritage and to the strong agricultural and forest products industries across the region. More frequent extreme heat episodes and changing seasonal climates are projected to increase exposure-linked health impacts and economic vulnerabilities in the agricultural, timber, and manufacturing sectors. By the end of the century, over one-half billion labor hours could be lost from extreme heat-related impacts. Such changes would negatively impact the region's labor-intensive agricultural industry and compound existing social stresses in rural areas related to limited local community capabilities and associated with rural demography, occupations, earnings, literacy, and poverty incidence. Reduction of existing stresses can increase resilience.

The Southeast includes vast expanses of coastal and inland low-lying areas, the southern portion of the Appalachian Mountains, numerous high-growth metropolitan areas, and large rural expanses. These beaches and bayous, fields and forests, and cities and small towns are all at risk from a changing climate. While some climate change impacts, such as sea level rise and extreme downpours, are being acutely felt now, others, like increasing exposure to dangerous high temperatures, humidity, and new local diseases, are expected to become more significant in the coming decades. While all regional residents and communities are potentially at risk for some impacts, some communities or populations are at greater risk due to their locations, services available to them, and economic situations.

Observed warming since the mid-20th century has been uneven in the Southeast region, with average daily minimum temperatures increasing three times faster than average daily maximum temperatures. The number of extreme rainfall events is increasing. Climate model simulations of future conditions project increases in both temperature and extreme precipitation.

Trends towards a more urbanized and denser Southeast are expected to continue, creating new climate vulnerabilities. Cities across the Southeast are experiencing more and longer summer heat waves. Vector-borne diseases pose a greater risk in cities than in rural areas because of higher population densities and other human factors, and the major urban centers in the Southeast are already impacted by poor air quality during warmer months. Increasing precipitation and extreme weather events will likely impact roads, freight rail, and passenger rail, which will likely have cascading effects across the region. Infrastructure related to drinking water and wastewater treatment also has the potential to be compromised by climate-related events. Increases in extreme rainfall events and high tide coastal floods due to future climate change will impact the quality of life of permanent residents as well as tourists visiting the low-lying and coastal regions of the Southeast. Sea level rise is contributing to increased coastal flooding in the Southeast, and high tide flooding already poses daily risks to businesses, neighborhoods, infrastructure, transportation, and ecosystems in the region. There have been numerous instances of intense rainfall events that have had devastating impacts on inland communities in recent years.

The ecological resources that people depend on for livelihoods, protection, and well-being are increasingly at risk from the impacts of climate change. Sea level rise will result in the rapid conversion of coastal, terrestrial, and freshwater ecosystems to tidal saline habitats. Reductions in the frequency and intensity of cold winter temperature extremes are already allowing tropical and subtropical species to move northward and replace more temperate species. Warmer winter temperatures are also expected to facilitate the northward

movement of problematic invasive species, which could transform natural systems north of their current distribution. In the future, rising temperatures and increases in the duration and intensity of drought are expected to increase wildfire occurrence and also reduce the effectiveness of prescribed fire practices.

Many in rural communities are maintaining connections to traditional livelihoods and relying on natural resources that are inherently vulnerable to climate changes. Climate trends and possible climate futures show patterns that are already impacting—and are projected to further impact—rural sectors, from agriculture and forestry to human health and labor productivity. Future temperature increases are projected to pose challenges to human health. Increases in temperatures, water stress, freeze-free days, drought, and wildfire risks, together with changing conditions for invasive species and the movement of diseases, create a number of potential risks for existing agricultural systems. Rural communities tend to be more vulnerable to these changes due to factors such as demography, occupations, earnings, literacy, and poverty incidence. In fact, a recent economic study using a higher scenario (RCP8.5) suggests that the southern and midwestern populations are likely to suffer the largest losses from future climate changes in the United States. Climate change tends to compound existing vulnerabilities and exacerbate existing inequities. Already poor regions, including those found in the Southeast, are expected to continue incurring greater losses than elsewhere in the United States.

For full chapter, including references and Traceable Accounts, see https://nca2018.globalchange.gov/chapter/southeast.

Historical Changes in Hot Days and Warm Nights

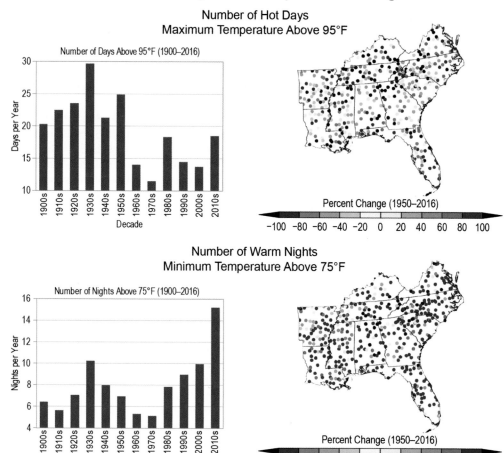

Number of Hot Days
Maximum Temperature Above 95°F

Number of Days Above 95°F (1900–2016)

Percent Change (1950–2016)

Number of Warm Nights
Minimum Temperature Above 75°F

Number of Nights Above 75°F (1900–2016)

Percent Change (1950–2016)

Sixty-one percent of major Southeast cities are exhibiting some aspects of worsening heat waves, which is a higher percentage than any other region of the country. Hot days and warm nights together impact human comfort and health and result in the need for increased cooling efforts. Agriculture is also impacted by a lack of nighttime cooling. Variability and change in (top) the annual number of hot days and (bottom) warm nights are shown. The bar charts show averages over the region by decade for 1900–2016, while the maps show the trends for 1950–2016 for individual weather stations. Average summer temperatures during the most recent 10 years have been the warmest on record, with very large increases in nighttime temperatures and more modest increases in daytime temperatures, as indicated by contrasting changes in hot days and warm nights. (top left) The annual number of hot days (maximum temperature above 95°F) has been lower since 1960 than the average during the first half of the 20th century; (top right) trends in hot days since 1950 are generally downward except along the south Atlantic coast and in Florida due to high numbers during the 1950s but have been slightly upward since 1960, following a gradual increase in average daytime maximum temperatures during that time. (bottom left) Conversely, the number of warm nights (minimum temperature above 75°F) has doubled on average compared to the first half of the 20th century and (bottom right) locally has increased at most stations. *From Figure 19.1 (Sources: NOAA NCEI and CICS-NC).*

Historical Change in Heavy Precipitation

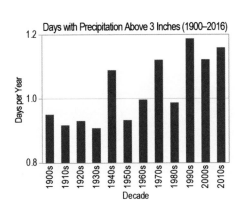

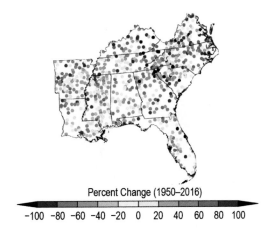

The figure shows variability and change in (left) the annual number of days with precipitation greater than 3 inches (1900–2016) averaged over the Southeast by decade and (right) individual station trends (1950–2016). The number of days with heavy precipitation has increased at most stations, particularly since the 1980s. *From Figure 19.3 (Sources: NOAA NCEI and CICS-NC).*

20 U.S. Caribbean

San Juan, Puerto Rico

Key Message 1

Freshwater

Freshwater is critical to life throughout the Caribbean. Increasing global carbon emissions are projected to reduce average rainfall in this region by the end of the century, constraining freshwater availability, while extreme rainfall events, which can increase freshwater flooding impacts, are expected to increase in intensity. Saltwater intrusion associated with sea level rise will reduce the quantity and quality of freshwater in coastal aquifers. Increasing variability in rainfall events and increasing temperatures will likely alter the distribution of ecological life zones and exacerbate existing problems in water management, planning, and infrastructure capacity.

Key Message 2

Marine Resources

Marine ecological systems provide key ecosystem services such as commercial and recreational fisheries and coastal protection. These systems are threatened by changes in ocean surface temperature, ocean acidification, sea level rise, and changes in the frequency and intensity of storm events. Degradation of coral and other marine habitats can result in changes in the distribution of species that use these habitats and the loss of live coral cover, sponges, and other key species. These changes will likely disrupt valuable ecosystem services, producing subsequent effects on Caribbean island economies.

Key Message 3

Coastal Systems

Coasts are a central feature of Caribbean island communities. Coastal zones dominate island economies and are home to critical infrastructure, public and private property, cultural heritage, and natural ecological systems. Sea level rise, combined with stronger wave action and higher storm surges, will worsen coastal flooding and increase coastal erosion, likely leading to diminished beach area, loss of storm surge barriers, decreased tourism, and negative effects on livelihoods and well-being. Adaptive planning and nature-based strategies, combined with active community participation and traditional knowledge, are beginning to be deployed to reduce the risks of a changing climate.

Key Message 4

Rising Temperatures

Natural and social systems adapt to the temperatures under which they evolve and operate. Changes to average and extreme temperatures have direct and indirect effects on organisms and strong interactions with hydrological cycles, resulting in a variety of impacts. Continued increases in average temperatures will likely lead to decreases in agricultural productivity, changes in habitats and wildlife distributions, and risks to human health, especially in vulnerable populations. As maximum and minimum temperatures increase, there are likely to be fewer cool nights and more frequent hot days, which will likely affect the quality of life in the U.S. Caribbean.

Key Message 5

Disaster Risk Response to Extreme Events

Extreme events pose significant risks to life, property, and economy in the Caribbean, and some extreme events, such as flooding and droughts, are projected to increase in frequency and intensity. Increasing hurricane intensity and associated rainfall rates will likely affect human health and well-being, economic development, conservation, and agricultural productivity. Increased resilience will depend on collaboration and integrated planning, preparation, and responses across the region.

Key Message 6

Increasing Adaptive Capacity Through Regional Collaboration

Shared knowledge, collaborative research and monitoring, and sustainable institutional adaptive capacity can help support and speed up disaster recovery, reduce loss of life, enhance food security, and improve economic opportunity in the U.S. Caribbean. Increased regional cooperation and stronger partnerships in the Caribbean can expand the region's collective ability to achieve effective actions that build climate change resilience, reduce vulnerability to extreme events, and assist in recovery efforts.

Historically, the U.S. Caribbean region has experienced relatively stable seasonal rainfall patterns, moderate annual temperature fluctuations, and a variety of extreme weather events, such as tropical storms, hurricanes, and drought. However, the Caribbean climate is changing and is projected to be increasingly variable as levels of greenhouse gases in the atmosphere increase.

The high percentage of coastal area relative to the total island land area in the U.S. Caribbean means that a large proportion of the region's people, infrastructure, and economic activity are vulnerable to sea level rise, more frequent intense rainfall events and associated coastal flooding, and saltwater intrusion. High levels of exposure and sensitivity to risk in the U.S. Caribbean region are compounded by a low level of adaptive capacity, due in part to the high costs of mitigation and adaptation measures relative to the region's gross domestic product, particularly when compared to continental U.S. coastal areas. The limited geographic and economic scale of Caribbean islands means that disruptions from extreme climate-related events, such as droughts and hurricanes, can devastate large portions of local economies and cause widespread damage to crops, water supplies, infrastructure, and other critical resources and services.

The U.S. Caribbean territories of Puerto Rico and the U.S. Virgin Islands (USVI) have distinct differences in topography, language, population size, governance, natural and human resources, and economic capacity. However, both are highly dependent on natural and built coastal assets; service-related industries account for more than 60% of the USVI economy. Beaches, affected by sea level rise and erosion, are among the main tourist attractions. In Puerto Rico, critical infrastructure (for example, drinking water pipelines and pump stations, sanitary pipelines and pump stations, wastewater treatment plants, and power plants) is vulnerable to the effects of sea level rise, storm surge, and flooding. In the USVI, infrastructure and historical buildings in the inundation zone for sea level rise include the power plants on both St. Thomas and St. Croix; schools; housing communities; the towns of Charlotte Amalie, Christiansted, and Frederiksted; and pipelines for water and sewage.

Climate change will likely result in water shortages due to an overall decrease in annual rainfall, a reduction in ecosystem services, and increased risks for agriculture, human health, wildlife, and socioeconomic development in the U.S. Caribbean. These shortages would result from some locations within the Caribbean experiencing longer dry seasons and shorter, but wetter, wet seasons in the future. Extended dry seasons are projected to increase fire likelihood. Excessive rainfall, coupled with poor construction practices, unpaved roads, and steep slopes, can exacerbate erosion rates and have adverse effects on reservoir capacity, water quality, and nearshore marine habitats.

Ocean warming poses a significant threat to the survival of corals and will likely also cause shifts in associated habitats that compose the coral reef ecosystem. Severe, repeated, or prolonged periods of high temperatures leading to extended coral bleaching can result in colony death. Ocean acidification also is likely to diminish the structural integrity of coral habitats. Studies show that major shifts in fisheries distribution and changes to the structure and composition of marine habitats adversely affect food security, shoreline protection, and economies throughout the Caribbean.

In Puerto Rico, the annual number of days with temperatures above 90°F has increased over the last four and a half decades. During that period, stroke and cardiovascular disease, which are influenced by such elevated temperatures, became the primary causes of death. Increases in average temperature and in extreme heat events will likely have detrimental effects on agricultural operations throughout the U.S. Caribbean region. Many farmers in the tropics, including the U.S. Caribbean, are considered small-holding, limited resource farmers and often lack the resources and/or capital to adapt to changing conditions.

Most Caribbean countries and territories share the need to assess risks, enable actions across scales, and assess changes in ecosystems to inform decision-making on habitat protection under a changing climate. U.S. Caribbean islands have the potential to improve adaptation and mitigation actions by fostering stronger collaborations with Caribbean initiatives on climate change and disaster risk reduction.

For full chapter, including references and Traceable Accounts, see https://nca2018. globalchange.gov/chapter/caribbean.

Observed and Projected Sea Level Rise

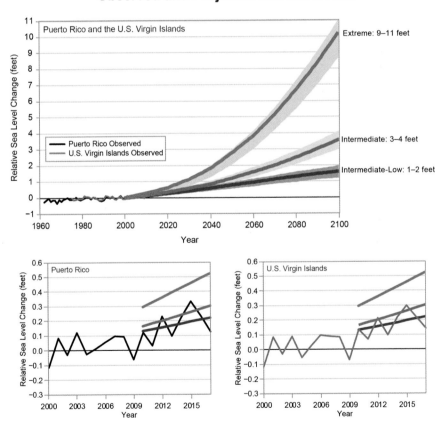

(top) Observed sea level rise trends in Puerto Rico and the U.S. Virgin Islands reflect an increase in sea level of about 0.08 inches (2.0 mm) per year for the period 1962–2017 for Puerto Rico and for 1975–2017 for the U.S. Virgin Islands. The bottom panels show a closer look at more recent trends from 2000 to 2017 that measure a rise in sea level of about 0.24 inches (6.0 mm) per year. Projections of sea level rise are shown under three different scenarios of Intermediate-Low (1–2 feet), Intermediate (3–4 feet), and Extreme (9–11 feet) sea level rise. The scenarios depict the range of future sea level rise based on factors such as global greenhouse gas emissions and the loss of glaciers and ice sheets. *From Figure 20.6 (Sources: NOAA NCEI and CICS-NC).*

Climate Risk Management Organizations

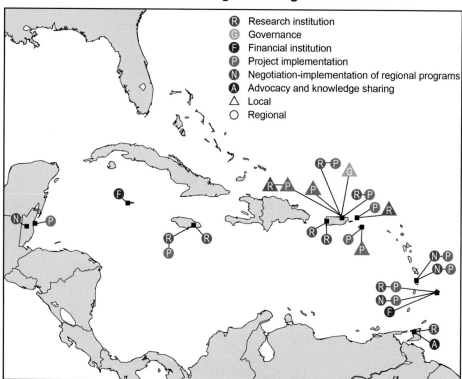

Some of the organizations working on climate risk assessment and management in the Caribbean are shown. Joint regional efforts to address climate challenges include the implementation of adaptation measures to reduce natural, social, and economic vulnerabilities, as well as actions to reduce greenhouse gas emissions. See the online version of this figure at http://nca2018.globalchange.gov/chapter/20#fig-20-18 for more details. *From Figure 20.18 (Sources: NOAA and the USDA Caribbean Climate Hub).*

21 Midwest

Carson, Wisconsin

Key Message 1

Agriculture

The Midwest is a major producer of a wide range of food and animal feed for national consumption and international trade. Increases in warm-season absolute humidity and precipitation have eroded soils, created favorable conditions for pests and pathogens, and degraded the quality of stored grain. Projected changes in precipitation, coupled with rising extreme temperatures before mid-century, will reduce Midwest agricultural productivity to levels of the 1980s without major technological advances.

Key Message 2

Forestry

Midwest forests provide numerous economic and ecological benefits, yet threats from a changing climate are interacting with existing stressors such as invasive species and pests to increase tree mortality and reduce forest productivity. Without adaptive actions, these interactions will result in the loss of economically and culturally important tree species such as paper birch and black ash and are expected to lead to the conversion of some forests to other forest types or even to non-forested ecosystems by the end of the century. Land managers are beginning to manage risk in forests by increasing diversity and selecting for tree species adapted to a range of projected conditions.

Key Message 3

Biodiversity and Ecosystems

The ecosystems of the Midwest support a diverse array of native species and provide people with essential services such as water purification, flood control, resource provision, crop pollination, and recreational opportunities. Species and ecosystems, including the important freshwater resources of the Great Lakes, are typically most at risk when climate stressors, like temperature increases, interact with land-use change, habitat loss, pollution, nutrient inputs, and nonnative invasive species. Restoration of natural systems, increases in the use of green infrastructure, and targeted conservation efforts, especially of wetland systems, can help protect people and nature from climate change impacts.

Key Message 4

Human Health

Climate change is expected to worsen existing health conditions and introduce new health threats by increasing the frequency and intensity of poor air quality days, extreme high temperature events, and heavy rainfalls; extending pollen seasons; and modifying the distribution of disease-carrying pests and insects. By mid-century, the region is projected to experience substantial, yet avoidable, loss of life, worsened health conditions, and economic impacts estimated in the billions of dollars as a result of these changes. Improved basic health services and increased public health measures—including surveillance and monitoring—can prevent or reduce these impacts.

Key Message 5

Transportation and Infrastructure

Storm water management systems, transportation networks, and other critical infrastructure are already experiencing impacts from changing precipitation patterns and elevated flood risks. Green infrastructure is reducing some of the negative impacts by using plants and open space to absorb storm water. The annual cost of adapting urban storm water systems to more frequent and severe storms is projected to exceed $500 million for the Midwest by the end of the century.

Key Message 6

Community Vulnerability and Adaptation

At-risk communities in the Midwest are becoming more vulnerable to climate change impacts such as flooding, drought, and increases in urban heat islands. Tribal nations are especially vulnerable because of their reliance on threatened natural resources for their cultural, subsistence, and economic needs. Integrating climate adaptation into planning processes offers an opportunity to better manage climate risks now. Developing knowledge for decision-making in cooperation with vulnerable communities and tribal nations will help to build adaptive capacity and increase resilience.

The Midwest is home to over 60 million people, and its active economy represents 18% of the U.S. gross domestic product. The region is probably best known for agricultural production. Increases in growing-season temperature in the Midwest are projected to be the largest contributing factor to declines in the productivity of U.S. agriculture. Increases in humidity in spring through mid-century are expected to increase rainfall, which will increase the potential for soil erosion and further reduce planting-season workdays due to waterlogged soil.

Forests are a defining characteristic of many landscapes within the Midwest, covering more than 91 million acres. However, a changing climate, including an increased frequency of late-growing-season drought conditions, is worsening the effects of invasive species, insect pests, and plant disease as trees experience periodic moisture stress. Impacts from human activities, such as logging, fire suppression, and agricultural expansion, have lowered the diversity of the Midwest's forests from the pre-Euro-American settlement period. Natural resource managers are taking steps to address these issues by increasing the diversity of trees and introducing species suitable for a changing climate.

The Great Lakes play a central role in the Midwest and provide an abundant freshwater resource for water supplies, industry, shipping, fishing, and recreation, as well as a rich and diverse ecosystem. These important ecosystems are under stress from pollution, nutrient and sediment inputs from agricultural systems, and invasive species. Lake surface temperatures are increasing, lake ice cover is declining, the seasonal stratification of temperatures in the lakes is occurring earlier in the year, and summer evaporation rates are increasing. Increasing storm impacts and declines in coastal water quality can put coastal communities at risk. While several coastal communities have expressed willingness to integrate climate action into planning efforts, access to useful climate information and limited human and financial resources constrain municipal action.

Land conversion, and a wide range of other stressors, has already greatly reduced biodiversity in many of the region's prairies, wetlands, forests, and freshwater systems. Species are already responding to changes that have occurred over the last several decades, and rapid climate change over the next century is expected to cause or further amplify stress in many species and ecological systems in the

Midwest. The loss of species and the degradation of ecosystems have the potential to reduce or eliminate essential ecological services such as flood control, water purification, and crop pollination, thus reducing the potential for society to successfully adapt to ongoing changes. However, understanding these relationships also highlights important climate adaptation strategies. For example, restoring systems like wetlands and forested floodplains and implementing agricultural best management strategies that increase vegetative cover (cover crops and riparian buffers) can help reduce flooding risks and protect water quality.

Midwestern populations are already experiencing adverse health impacts from climate change, and these impacts are expected to worsen in the future. In the absence of mitigation, ground-level ozone concentrations are projected to increase across most of the Midwest, resulting in an additional 200–550

premature deaths in the region per year by 2050. Exposure to high temperatures impacts workers' health, safety, and productivity. Currently, days over 100°F in Chicago are rare. However, they could become increasingly more common by late century in both the lower and higher scenarios (RCP4.5 and RCP8.5).

The Midwest also has vibrant manufacturing, retail, recreation/tourism, and service sectors. The region's highways, railroads, airports, and navigable rivers are major modes for commerce activity. Increasing precipitation, especially heavy rain events, has increased the overall flood risk, causing disruption to transportation and damage to property and infrastructure. Increasing use of green infrastructure (including nature-based approaches, such as wetland restoration, and innovations like permeable pavements) and better engineering practices are beginning to address these issues.

Conservation Practices Reduce Impact of Heavy Rains
Integrating strips of native prairie vegetation into row crops has been shown to reduce sediment and nutrient loss from fields, as well as improve biodiversity and the delivery of ecosystem services. Iowa State University's STRIPS program is actively conducting research into this agricultural conservation practice. The inset shows a close-up example of a prairie vegetation strip. *From Figure 21.2 (Photo credits: [main photo] Lynn Betts, [inset] Farnaz Kordbacheh).*

Citizens and stakeholders value their health and the well-being of their communities—all of which are at risk from increased flooding, increased heat, and lower air and water quality under a changing climate. To better prevent and respond to these impacts, scholars and practitioners highlight the need to engage in risk-driven approaches that not only focus on assessing vulnerabilities but also include effective planning and implementation of adaptation options.

For full chapter, including references and Traceable Accounts, see https://nca2018. globalchange.gov/chapter/midwest.

Forest Diversity Can Increase Resilience to Climate Change
The photo shows Menominee Tribal Enterprises staff creating opportunity from adversity by replanting a forest opening caused by oak wilt disease with a diverse array of tree and understory plant species that are expected to fare better under future climate conditions. *From Figure 21.4 (Photo credit: Kristen Schmitt).*

22 Northern Great Plains

Cattle grazing in the plains of western Montana

Key Message 1

Water

Water is the lifeblood of the Northern Great Plains, and effective water management is critical to the region's people, crops and livestock, ecosystems, and energy industry. Even small changes in annual precipitation can have large effects downstream; when coupled with the variability from extreme events, these changes make managing these resources a challenge. Future changes in precipitation patterns, warmer temperatures, and the potential for more extreme rainfall events are very likely to exacerbate these challenges.

Key Message 2

Agriculture

Agriculture is an integral component of the economy, the history, and the culture of the Northern Great Plains. Recently, agriculture has benefited from longer growing seasons and other recent climatic changes. Some additional production and conservation benefits are expected in the next two to three decades as land managers employ innovative adaptation strategies, but rising temperatures and changes in extreme weather events are very likely to have negative impacts on parts of the region. Adaptation to extremes and to longer-term, persistent climate changes will likely require transformative changes in agricultural management, including regional shifts of agricultural practices and enterprises.

Key Message 3

Recreation and Tourism

Ecosystems across the Northern Great Plains provide recreational opportunities and other valuable goods and services that are at risk in a changing climate. Rising temperatures have already resulted in shorter snow seasons, lower summer streamflows, and higher stream temperatures and have negatively affected high-elevation ecosystems and riparian areas, with important consequences for local economies that depend on winter or river-based recreational activities. Climate-induced land-use changes in agriculture can have cascading effects on closely entwined natural ecosystems, such as wetlands, and the diverse species and recreational amenities they support. Federal, tribal, state, and private organizations are undertaking preparedness and adaptation activities, such as scenario planning, transboundary collaboration, and development of market-based tools.

Key Message 4

Energy

Fossil fuel and renewable energy production and distribution infrastructure is expanding within the Northern Great Plains. Climate change and extreme weather events put this infrastructure at risk, as well as the supply of energy it contributes to support individuals, communities, and the U.S. economy as a whole. The energy sector is also a significant source of greenhouse gases and volatile organic compounds that contribute to climate change and ground-level ozone pollution.

Key Message 5

Indigenous Peoples

Indigenous peoples of the Northern Great Plains are at high risk from a variety of climate change impacts, especially those resulting from hydrological changes, including changes in snowpack, seasonality and timing of precipitation events, and extreme flooding and droughts as well as melting glaciers and reduction in streamflows. These changes are already resulting in harmful impacts to tribal economies, livelihoods, and sacred waters and plants used for ceremonies, medicine, and subsistence. At the same time, many tribes have been very proactive in adaptation and strategic climate change planning.

In the Northern Great Plains, the timing and quantity of both precipitation and runoff have important consequences for water supplies, agricultural activities, and energy production. Overall, climate projections suggest that the number of heavy precipitation events (events with greater than 1 inch per day of rainfall) is projected to increase. Moving forward, the magnitude of year-to-year variability overshadows the small projected average decrease in streamflow. Changes in extreme events are likely to overwhelm average changes in both the eastern and western regions of the Northern Great Plains. Major flooding across the basin in 2011 was followed by severe drought in 2012, representing new and unprecedented variability that is likely to become more common in a warmer world.

The Northern Great Plains region plays a critical role in national food security. Among other anticipated changes, projected warmer and generally wetter conditions with elevated atmospheric carbon dioxide concentrations are expected to increase the abundance and competitive ability of weeds and invasive species, increase livestock production and efficiency of production, and result in longer growing seasons at mid- and high latitudes. Net primary productivity, including crop yields and forage production, is also likely to increase, although an increasing number of extreme temperature events during critical pollination and grain fill periods is likely to reduce crop yields.

Ecosystems across the Northern Great Plains provide recreational opportunities and other valuable goods and services that are ingrained in the region's cultures. Higher temperatures, reduced snow cover, and more variable precipitation will make it increasingly challenging to manage the region's valuable wetlands, rivers, and snow-dependent ecosystems. In the mountains of western Wyoming and western Montana, the fraction of total water in precipitation that falls as snow is expected to decline by 25% to 40% by 2100 under a higher scenario (RCP8.5), which would negatively affect the region's winter recreation industry. At lower-elevation areas of the Northern Great Plains, climate-induced land-use changes in agriculture can have cascading effects on closely entwined natural ecosystems, such as wetlands, and the diverse species and recreational opportunities they support.

Energy resources in the Northern Great Plains include abundant crude oil, natural gas, coal, wind, and stored water, and to a lesser extent, corn-based ethanol, solar energy, and uranium. The infrastructure associated with the extraction, distribution, and energy produced from these resources is vulnerable to the impacts of climate change. Railroads and pipelines are vulnerable to damage or disruption from increasing heavy precipitation events and associated flooding and erosion. Declining water availability in the summer would likely increase costs for oil production operations, which require freshwater resources. These cost increases will either lead to lower production or be passed on to consumers. Finally, higher maximum temperatures, longer and more severe heat waves, and higher overnight lows are expected to increase electricity demand for cooling in the summer, further stressing the power grid.

Indigenous peoples in the region are observing changes to climate, many of which are impacting livelihoods as well as traditional subsistence and wild foods, wildlife, plants and water for ceremonies, medicines, and health and well-being. Because some tribes and Indigenous peoples are among those in the region with the highest rates of poverty and unemployment, and because many are still directly reliant on natural resources, they are among the most at risk to climate change.

For full chapter, including references and Traceable Accounts, see https://nca2018.global-change.gov/chapter/northern-great-plains.

Projected Changes in Very Hot Days, Cool Days, and Heavy Precipitation

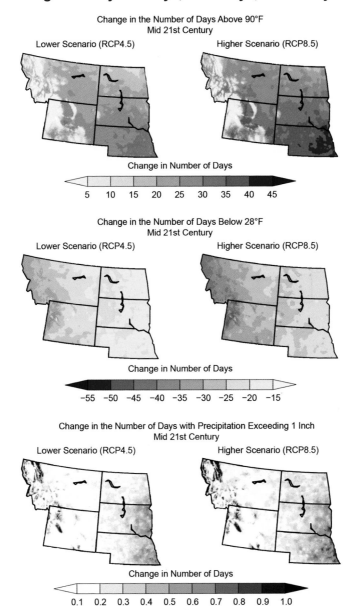

Projected changes are shown for (top) the annual number of very hot days (days with maximum temperatures above 90°F, an indicator of crop stress and impacts on human health), (middle) the annual number of cool days (days with minimum temperatures below 28°F, an indicator of damaging frost), and (bottom) heavy precipitation events (the annual number of days with greater than 1 inch of rainfall; areas in white do not normally experience more than 1 inch of rainfall in a single day). Projections are shown for the middle of the 21st century (2036–2065) as compared to the 1976–2005 average under the lower and higher scenarios (RCP4.5 and RCP8.5). *From Figure 22.2 (Sources: NOAA NCEI and CICS-NC).*

Northern Great Plains Tribal Lands

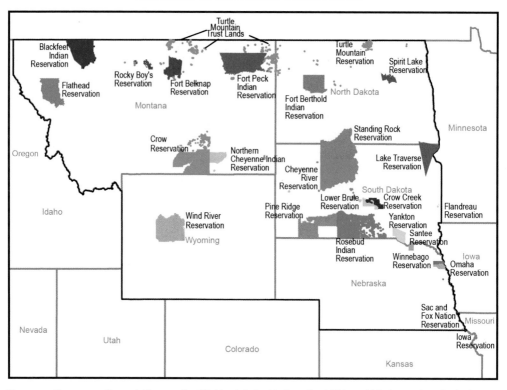

The map outlines reservation and off-reservation tribal lands in the Northern Great Plains, which shows where the 27 federally recognized tribes have a significant portion of lands throughout the region. Information on Indigenous peoples' climate projects within the Northern Great Plains is described in Chapter 15: Tribes and Indigenous Peoples. *From Figure 22.7 (Sources: created by North Central Climate Science Center [2017] with data from the Bureau of Indian Affairs, Colorado State University, and USGS National Map).*

23 | Southern Great Plains

Whooping cranes in the Aransas National Wildlife Refuge in Texas

Key Message 1

Food, Energy, and Water Resources

Quality of life in the region will be compromised as increasing population, the migration of individuals from rural to urban locations, and a changing climate redistribute demand at the intersection of food consumption, energy production, and water resources. A growing number of adaptation strategies, improved climate services, and early warning decision support systems will more effectively manage the complex regional, national, and transnational issues associated with food, energy, and water.

Key Message 2

Infrastructure

The built environment is vulnerable to increasing temperature, extreme precipitation, and continued sea level rise, particularly as infrastructure ages and populations shift to urban centers. Along the Texas Gulf Coast, relative sea level rise of twice the global average will put coastal infrastructure at risk. Regional adaptation efforts that harden or relocate critical infrastructure will reduce the risk of climate change impacts.

Key Message 3

Ecosystems and Ecosystem Services

Terrestrial and aquatic ecosystems are being directly and indirectly altered by climate change. Some species can adapt to extreme droughts, unprecedented floods, and wildfires from a changing climate, while others cannot, resulting in significant impacts to both services and people living in these ecosystems. Landscape-scale ecological services will increase the resilience of the most vulnerable species.

Key Message 4

Human Health

Health threats, including heat illness and diseases transmitted through food, water, and insects, will increase as temperature rises. Weather conditions supporting these health threats are projected to be of longer duration or occur at times of the year when these threats are not normally experienced. Extreme weather events with resultant physical injury and population displacement are also a threat. These threats are likely to increase in frequency and distribution and are likely to create significant economic burdens. Vulnerability and adaptation assessments, comprehensive response plans, seasonal health forecasts, and early warning systems can be useful adaptation strategies.

Key Message 5

Indigenous Peoples

Tribal and Indigenous communities are particularly vulnerable to climate change due to water resource constraints, extreme weather events, higher temperature, and other likely public health issues. Efforts to build community resilience can be hindered by economic, political, and infrastructure limitations, but traditional knowledge and intertribal organizations provide opportunities to adapt to the potential challenges of climate change.

The Southern Great Plains experiences weather that is dramatic and consequential; from hurricanes and flooding to heat waves and drought, its 34 million people, their infrastructure, and economies are often stressed, greatly impacting socioeconomic systems. The quality of life for the region's residents is dependent upon resources and natural systems for the sustainable provision of our basic needs—food, energy, and water. Extreme weather and climate events have redistributed demands for consumption, production, and supply across the region. Adaptation strategies that integrate climate services and early warning systems are improving our abilities to develop sustainable infrastructure and increase agricultural production, yet include the flexibility needed to embrace any changing demand patterns.

Regional adaptation efforts that harden or relocate critical infrastructure will reduce the risk of climate change impacts. Redesigns of coastal infrastructure and the use of green/gray methodologies are improving future coastal resilience. Energy industry reinvention is ensuring operations and reliability during extreme climatic events. Increasingly robust considerations of economic resilience allow us to anticipate risk, evaluate how that risk can affect our needs, and build a responsive adaptive capacity.

With climate change, terrestrial and aquatic ecosystems, and species within them, have winners and losers. Those that can adapt are "increasers," while others cannot, resulting in impacts to traditional services and the livelihoods of the people who depend on those resources. The warming of coastal bay waters has been documented since at least the 1980s, and those increases in water temperature directly affect water quality, leading to hypoxia, harmful algal blooms, and fish kills—thus lowering the productivity and diversity of estuaries. Natural wetlands like the playa lakes in the High Plains, which have served for centuries as important habitat for migrating waterfowl, are virtually nonexistent during drought.

Direct human health threats follow a similar pattern of species within our natural ecosystems. Extreme weather results in both direct and indirect impacts to people; physical injury and population displacement are anticipated to result with climate change. Heat illness and diseases transmitted through food, water, and insects increase human risk as temperature rises. Acute awareness of these future impacts allows us to plan for the most vulnerable and adapt through response plans, health forecasting, and early warning strategies, including those that span transboundary contexts and systems.

The impacts of climate change in general become more acute when considering tribal and Indigenous communities. Resilience to climate change will be hindered by economic, political, and infrastructure limitations for these groups;

at the same time, connectivity of the tribes and Indigenous communities offers opportunities for teaching adaptably through their cultural means of applying traditional knowledge and intertribal organization. These well-honed connections of adapting through the centuries may help all of us learn how to offset the impacts and potential challenges of climate change.

The role of climate change in altering the frequency of the types of severe weather most typically associated with the Southern Great Plains, such as severe local storms, hailstorms, and tornadoes, remains difficult to quantify. Indirect approaches suggest a possible increase in the circumstances conducive to such severe weather, including an increase in the instances of larger hail sizes in the region by 2040, but changes are unlikely to be uniform across the region, and additional research is needed.

For full chapter, including references and Traceable Accounts, see https://nca2018. globalchange.gov/chapter/southern-great-plains.

Projected Increase in Number of Days Above 100°F

Late 21st Century

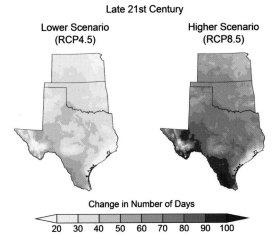

Lower Scenario
(RCP4.5)

Higher Scenario
(RCP8.5)

Change in Number of Days

20 30 40 50 60 70 80 90 100

Under both lower- and higher-scenario climate change projections, the number of days exceeding 100°F is projected to increase markedly across the Southern Great Plains by the end of the century (2070–2099 as compared to 1976–2005). *From Figure 23.4 (Sources: NOAA NCEI and CICS-NC).*

24 | Northwest

Four Lakes basin in White Cloud Peaks, Sawtooth National Forest, Idaho

Key Message 1

Natural Resource Economy

Climate change is already affecting the Northwest's diverse natural resources, which support sustainable livelihoods; provide a robust foundation for rural, tribal, and Indigenous communities; and strengthen local economies. Climate change is expected to continue affecting the natural resource sector, but the economic consequences will depend on future market dynamics, management actions, and adaptation efforts. Proactive management can increase the resilience of many natural resources and their associated economies.

Key Message 2

Natural World and Cultural Heritage

Climate change and extreme events are already endangering the well-being of a wide range of wildlife, fish, and plants, which are intimately tied to tribal subsistence culture and popular outdoor recreation activities. Climate change is projected to continue to have adverse impacts on the regional environment, with implications for the values, identity, heritage, cultures, and quality of life of the region's diverse population. Adaptation and informed management, especially culturally appropriate strategies, will likely increase the resilience of the region's natural capital.

Key Message 3

Infrastructure

Existing water, transportation, and energy infrastructure already face challenges from flooding, landslides, drought, wildfire, and heat waves. Climate change is projected to increase the risks from many of these extreme events, potentially compromising the reliability of water supplies, hydropower, and transportation across the region. Isolated communities and those with systems that lack redundancy are the most vulnerable. Adaptation strategies that address more than one sector, or are coupled with social and environmental co-benefits, can increase resilience.

Key Message 4

Health

Organizations and volunteers that make up the Northwest's social safety net are already stretched thin with current demands. Healthcare and social systems will likely be further challenged with the increasing frequency of acute events, or when cascading events occur. In addition to an increased likelihood of hazards and epidemics, disruptions in local economies and food systems are projected to result in more chronic health risks. The potential health co-benefits of future climate mitigation investments could help to counterbalance these risks.

Key Message 5

Frontline Communities

Communities on the front lines of climate change experience the first, and often the worst, effects. Frontline communities in the Northwest include tribes and Indigenous peoples, those most dependent on natural resources for their livelihoods, and the economically disadvantaged. These communities generally prioritize basic needs, such as shelter, food, and transportation; frequently lack economic and political capital; and have fewer resources to prepare for and cope with climate disruptions. The social and cultural cohesion inherent in many of these communities provides a foundation for building community capacity and increasing resilience.

Residents of the Northwest list the inherent qualities of the natural environment among the top reasons to live in the region. The region is known for clean air, abundant water, low-cost hydroelectric power, vast forests, extensive farmlands, and outdoor recreation that includes hiking, boating, fishing, hunting, and skiing. Climate change, including gradual changes to the climate and in extreme climatic events, is already affecting these valued aspects of the region, including the natural resource sector, cultural identity and quality of life, built infrastructure systems, and the health of Northwest residents. The communities on the front lines of climate change—tribes and Indigenous peoples, those most dependent on natural resources for their livelihoods, and the economically disadvantaged—are experiencing the first, and often the worst, effects.

In the Third National Climate Assessment, the Key Messages for the Northwest focused on projected climate impacts to the region. These impacts, many of which are now better understood in the scientific literature, remain the primary climate concerns over the coming decades. In this updated assessment, the Key Messages explore how climate change could affect the interrelationships between the environment and the people of the Northwest. The extreme weather events of 2015 provide an excellent opportunity to explore projected changes in baseline climate conditions for the Northwest. The vast array of climate impacts that occurred over this record-breaking warm and dry year, coupled with the impacts of a multiyear drought, provide an enlightening glimpse into what may be more commonplace under a warmer future climate. Record-low snowpack led to water scarcity and large wildfires that negatively affected farmers, hydropower, drinking water, air quality, salmon, and recreation. Warmer than normal ocean temperatures led to shifts in the marine ecosystem, challenges for salmon, and a large harmful algal bloom that adversely affected the region's fisheries and shellfish harvests.

Strong climate variability is likely to persist for the Northwest, owing in part to the year-to-year and decade-to-decade climate variability associated with the Pacific Ocean. Periods of prolonged drought are projected to be interspersed with years featuring heavy rainfall driven by powerful atmospheric rivers and strong El Niño winters associated with storm surge, large waves, and coastal erosion. Continued changes in the ocean environment, such as warmer waters, altered chemistry, sea level rise, and shifts in the marine ecosystems are also expected. These changes would affect the Northwest's natural resource economy, cultural heritage, built infrastructure, and recreation as well as the health and welfare of Northwest residents.

The Northwest has an abundance of examples and case studies that highlight climate adaptation in progress and in practice—including creating resilient agro-ecosystems that reduce climate-related risks while meeting economic, conservation, and adaptation goals; using "green" or hybrid "green and gray" infrastructure solutions that combine nature-based solutions with more traditional engineering approaches; and building social cohesion and strengthening social networks in frontline communities to assist in meeting basic needs while also increasing resilience to future climate stressors. Many of the case studies in this chapter demonstrate the importance of co-producing adaptation efforts with scientists, resource managers, communities, and decision-makers as the region prepares for climate change impacts across multiple sectors and resources.

For full chapter, including references and Traceable Accounts, see https://nca2018.globalchange.gov/chapter/northwest.

Climate Change Will Impact Key Aspects of Life in the Northwest

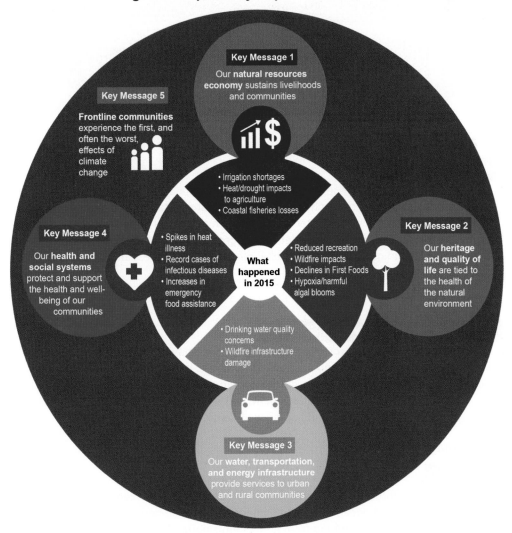

The climate-related events of 2015 provide a glimpse into the Northwest's future, because the kinds of extreme events that affected the Northwest in 2015 are projected to become more common. The climate impacts that occurred during this record-breaking warm and dry year highlight the close interrelationships between the climate, the natural and built environment, and the health and well-being of the Northwest's residents. *From Figure 24.2 (Source: USGCRP).*

25 Southwest

Low water levels in Lake Mead

Key Message 1

Water Resources

Water for people and nature in the Southwest has declined during droughts, due in part to human-caused climate change. Intensifying droughts and occasional large floods, combined with critical water demands from a growing population, deteriorating infrastructure, and groundwater depletion, suggest the need for flexible water management techniques that address changing risks over time, balancing declining supplies with greater demands.

Key Message 2

Ecosystems and Ecosystem Services

The integrity of Southwest forests and other ecosystems and their ability to provide natural habitat, clean water, and economic livelihoods have declined as a result of recent droughts and wildfire due in part to human-caused climate change. Greenhouse gas emissions reductions, fire management, and other actions can help reduce future vulnerabilities of ecosystems and human well-being.

Key Message 3

The Coast

Many coastal resources in the Southwest have been affected by sea level rise, ocean warming, and reduced ocean oxygen—all impacts of human-caused climate change—and ocean acidification resulting from human emissions of carbon dioxide. Homes and other coastal infrastructure, marine flora and fauna, and people who depend on coastal resources face increased risks under continued climate change.

Key Message 4

Indigenous Peoples

Traditional foods, natural resource-based livelihoods, cultural resources, and spiritual well-being of Indigenous peoples in the Southwest are increasingly affected by drought, wildfire, and changing ocean conditions. Because future changes would further disrupt the ecosystems on which Indigenous peoples depend, tribes are implementing adaptation measures and emissions reduction actions.

Key Message 5

Energy

The ability of hydropower and fossil fuel electricity generation to meet growing energy use in the Southwest is decreasing as a result of drought and rising temperatures. Many renewable energy sources offer increased electricity reliability, lower water intensity of energy generation, reduced greenhouse gas emissions, and new economic opportunities.

Key Message 6

Food

Food production in the Southwest is vulnerable to water shortages. Increased drought, heat waves, and reduction of winter chill hours can harm crops and livestock; exacerbate competition for water among agriculture, energy generation, and municipal uses; and increase future food insecurity.

Key Message 7

Human Health

Heat-associated deaths and illnesses, vulnerabilities to chronic disease, and other health risks to people in the Southwest result from increases in extreme heat, poor air quality, and conditions that foster pathogen growth and spread. Improving public health systems, community infrastructure, and personal health can reduce serious health risks under future climate change.

The Southwest region encompasses diverse ecosystems, cultures, and economies, reflecting a broad range of climate conditions, including the hottest and driest climate in the United States. Water for people and nature in the Southwest region has declined during droughts, due in part to human-caused climate change. Higher temperatures intensified the recent severe drought in California and are amplifying drought in the Colorado River Basin. Since 2000, Lake Mead on the Colorado River has fallen 130 feet (40 m) and lost 60% of its volume, a result of the ongoing Colorado River Basin drought and continued water withdrawals by cities and agriculture.

The reduction of water volume in both Lake Powell and Lake Mead increases the risk of water shortages across much of the Southwest. Local water utilities, the governments of seven U.S. states, and the federal governments of the United States and Mexico have voluntarily developed and implemented solutions to minimize the possibility of water shortages for cities, farms, and ecosystems. In response to the recent California drought, the state implemented a water conservation plan in 2014 that set allocations for water utilities and major users and banned wasteful practices. As a result, the people of the state reduced water use 25% from 2014 to 2017.

Exposure to hotter temperatures and heat waves already leads to heat-associated deaths in Arizona and California. Mortality risk during a heat wave is amplified on days with high levels of ground-level ozone or particulate air pollution. Given the proportion of the U.S. population in the Southwest region, a disproportionate number of West Nile virus, plague, hantavirus pulmonary syndrome, and Valley fever cases occur in the region.

Analyses estimated that the area burned by wildfire across the western United States from 1984 to 2015 was twice what would have burned had climate change not occurred. Wildfires around Los Angeles from 1990 to 2009 caused $3.1 billion in damages (unadjusted for inflation). Tree death in mid-elevation conifer forests doubled from 1955 to 2007 due, in part, to climate change. Allowing naturally ignited fires to burn in wilderness areas and preemptively setting low-severity prescribed burns in areas of unnatural fuel accumulations can reduce the risk of high-severity fires under climate change. Reducing greenhouse gas emissions globally can also reduce ecological vulnerabilities.

At the Golden Gate Bridge in San Francisco, sea level rose 9 inches (22 cm) between 1854 and 2016. Climate change caused most of this rise by melting of land ice and thermal expansion of ocean water. Local governments on the California coast are using projections of sea level rise to develop plans to reduce future risks. Ocean water acidity off the coast of California increased 25% to 40% (decreases of 0.10 to 0.15 pH units) from the preindustrial era (circa 1750) to 2014 due to increasing concentrations of atmospheric carbon dioxide from human activities. The marine heat wave along the Pacific Coast from 2014 to 2016 occurred due to a combination of natural factors and climate change. The event led to the mass stranding of sick and starving birds and sea lions, and shifts of red crabs and tuna into the region. The ecosystem disruptions contributed to closures of commercially important fisheries.

Agricultural irrigation accounts for approximately three-quarters of water use in the Southwest region, which grows half of the fruits, vegetables, and nuts and most of the wine grapes, strawberries, and lettuce for the

United States. Increasing heat stress during specific phases of the plant life cycle can increase crop failures.

Drought and increasing heat intensify the arid conditions of reservations where the United States restricted some tribal nations in the Southwest region to the driest portions of their traditional homelands. In response to climate change, Indigenous peoples in the region are developing new adaptation and mitigation actions.

The severe drought in California, intensified by climate change, reduced hydroelectric generation two-thirds from 2011 to 2015.

The efficiency of all water-cooled electric power plants that burn fuel depends on the temperature of the external cooling water, so climate change could reduce energy efficiency up to 15% across the Southwest by 2050. Solar, wind, and other renewable energy sources, except biofuels, emit less carbon and require less water than fossil fuel energy. Economic conditions and technological innovations have lowered renewable energy costs and increased renewable energy generation in the Southwest.

For full chapter, including references and Traceable Accounts, see https://nca2018.globalchange.gov/chapter/southwest.

Climate Change Has Increased Wildfire

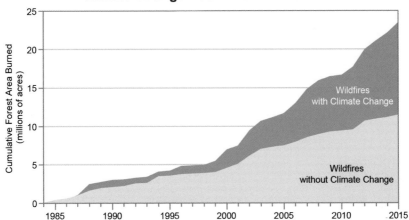

The cumulative forest area burned by wildfires has greatly increased between 1984 and 2015, with analyses estimating that the area burned by wildfire across the western United States over that period was twice what would have burned had climate change not occurred. *From Figure 25.4 (Source: adapted from Abatzoglou and Williams 2016).*

Fourth National Climate Assessment

Severe Drought Reduces Water Supplies in the Southwest

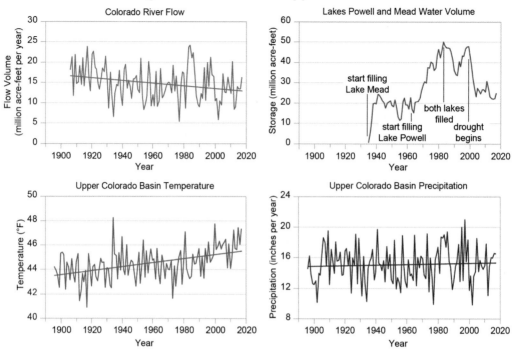

Since 2000, drought that was intensified by long-term trends of higher temperatures due to climate change has reduced the flow in the Colorado River (top left), which in turn has reduced the combined contents of Lakes Powell and Mead to the lowest level since both lakes were first filled (top right). In the Upper Colorado River Basin that feeds the reservoirs, temperatures have increased (bottom left), which increases plant water use and evaporation, reducing lake inflows and contents. Although annual precipitation (bottom right) has been variable without a long-term trend, there has been a recent decline in precipitation that exacerbates the drought. Combined with increased Lower Basin water consumption that began in the 1990s, these trends explain the recently reduced reservoir contents. Straight lines indicate trends for temperature, precipitation, and river flow. The trends for temperature and river flow are statistically significant. *From Figure 25.3 (Sources: Colorado State University and CICS-NC. Temperature and precipitation data from: PRISM Climate Group, Oregon State University, http://prism.oregonstate.edu, accessed 20 Jun 2018).*

26 Alaska

Anchorage, Alaska

Key Message 1

Marine Ecosystems

Alaska's marine fish and wildlife habitats, species distributions, and food webs, all of which are important to Alaska's residents, are increasingly affected by retreating and thinning arctic summer sea ice, increasing temperatures, and ocean acidification. Continued warming will accelerate related ecosystem alterations in ways that are difficult to predict, making adaptation more challenging.

Key Message 2

Terrestrial Processes

Alaska residents, communities, and their infrastructure continue to be affected by permafrost thaw, coastal and river erosion, increasing wildfire, and glacier melt. These changes are expected to continue into the future with increasing temperatures, which would directly impact how and where many Alaskans will live.

Key Message 3

Human Health

A warming climate brings a wide range of human health threats to Alaskans, including increased injuries, smoke inhalation, damage to vital water and sanitation systems, decreased food and water security, and new infectious diseases. The threats are greatest for rural residents, especially those who face increased risk of storm damage and flooding, loss of vital food sources, disrupted traditional practices, or relocation. Implementing adaptation strategies would reduce the physical, social, and psychological harm likely to occur under a warming climate.

Key Message 4

Indigenous Peoples

The subsistence activities, culture, health, and infrastructure of Alaska's Indigenous peoples and communities are subject to a variety of impacts, many of which are expected to increase in the future. Flexible, community-driven adaptation strategies would lessen these impacts by ensuring that climate risks are considered in the full context of the existing sociocultural systems.

Key Message 5

Economic Costs

Climate warming is causing damage to infrastructure that will be costly to repair or replace, especially in remote Alaska. It is also reducing heating costs throughout the state. These effects are very likely to grow with continued warming. Timely repair and maintenance of infrastructure can reduce the damages and avoid some of these added costs.

Key Message 6

Adaptation

Proactive adaptation in Alaska would reduce both short- and long-term costs associated with climate change, generate social and economic opportunity, and improve livelihood security. Direct engagement and partnership with communities is a vital element of adaptation in Alaska.

Alaska is the largest state in the Nation, almost one-fifth the size of the combined lower 48 United States, and is rich in natural capital resources. Alaska is often identified as being on the front lines of climate change since it is warming faster than any other state and faces a myriad of issues associated with a changing climate. The cost of infrastructure damage from a warming climate is projected to be very large, potentially ranging from $110 to $270 million per year, assuming timely repair and maintenance.

Although climate change does and will continue to dramatically transform the climate and environment of the Arctic, proactive adaptation in Alaska has the potential to reduce costs associated with these impacts. This includes the dissemination of several tools, such as guidebooks to support adaptation planning, some of which focus on Indigenous communities. While many opportunities exist with a changing climate, economic prospects are not well captured in the literature at this time.

As the climate continues to warm, there is likely to be a nearly sea ice-free Arctic during the summer by mid-century. Ocean acidification is an emerging global problem that will intensify with continued carbon dioxide (CO_2) emissions and negatively affects organisms. Climate change will likely affect management actions and economic drivers, including fisheries, in complex ways. The use of multiple alternative models to appropriately characterize uncertainty in future fisheries biomass trajectories and harvests could help manage these challenges. As temperature and precipitation increase across the Alaska landscape, physical and biological changes are also occurring throughout Alaska's terrestrial ecosystems. Degradation of permafrost is expected to continue, with associated impacts to infrastructure, river and stream discharge, water quality, and fish and wildlife habitat.

Longer sea ice-free seasons, higher ground temperatures, and relative sea level rise are expected to exacerbate flooding and accelerate erosion in many regions, leading to the loss of terrestrial habitat in the future and in some cases requiring entire communities or portions of communities to relocate to safer terrain. The influence of climate change on human health in Alaska can be traced to three sources: direct exposures, indirect effects, and social or psychological disruption. Each of these will have different manifestations for Alaskans when compared to residents elsewhere in the United States. Climate change exerts indirect effects on human health in Alaska through changes to water, air, and soil and through ecosystem changes affecting disease ecology and food security, especially in rural communities.

Alaska's rural communities are predominantly inhabited by Indigenous peoples who may be disproportionately vulnerable to socioeconomic and environmental change; however, they also have rich cultural traditions of resilience and adaptation. The impacts of climate change will likely affect all aspects of Alaska Native societies, from nutrition, infrastructure, economics, and health consequences to language, education, and the communities themselves.

The profound and diverse climate-driven changes in Alaska's physical environment and ecosystems generate economic impacts through their effects on environmental services. These services include positive benefits directly from ecosystems (for example, food, water, and other resources), as well as services provided directly from the physical environment (for example, temperature moderation, stable ground for supporting infrastructure, and smooth surface for overland transportation). Some of these effects are relatively assured and in some cases are already occurring. Other impacts are highly uncertain, due to their dependence on the structure of global and regional economies and future human alterations to the environment decades into the future, but they could be large.

In Alaska, a range of adaptations to changing climate and related environmental conditions are underway and others have been proposed as potential actions, including measures to reduce vulnerability and risk, as well as more systemic institutional transformation.

For full chapter, including references and Traceable Accounts, see https://nca2018. globalchange.gov/chapter/alaska.

Adaptation Planning in Alaska

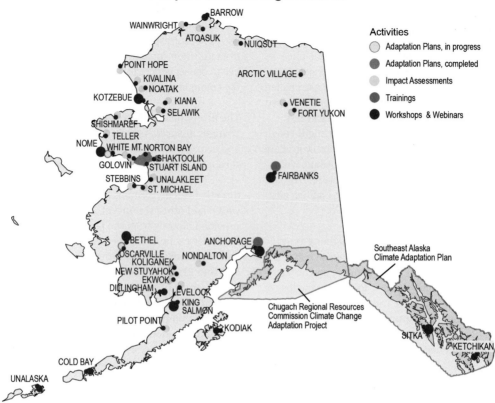

The map shows tribal climate adaptation planning efforts in Alaska. Research is considered to be adaptation under some classification schemes. Alaska is scientifically data poor, compared to other Arctic regions. In addition to research conducted at universities and by federal scientists, local community observer programs exist through several organizations, including the National Weather Service for weather and river ice observations; the University of Alaska for invasive species; and the Alaska Native Tribal Health Consortium for local observations of environmental change. Additional examples of community-based monitoring can be found through the website of the Alaska Ocean Observing System. *From Figure 26.9 (Source: adapted from Meeker and Kettle 2017).*

27 Hawai'i and U.S.-Affiliated Pacific Islands

Honolulu, Hawai'i

Key Message 1

Threats to Water Supplies

Dependable and safe water supplies for Pacific island communities and ecosystems are threatened by rising temperatures, changing rainfall patterns, sea level rise, and increased risk of extreme drought and flooding. Islands are already experiencing saltwater contamination due to sea level rise, which is expected to catastrophically impact food and water security, especially on low-lying atolls. Resilience to future threats relies on active monitoring and management of watersheds and freshwater systems.

Key Message 2

Terrestrial Ecosystems, Ecosystem Services, and Biodiversity

Pacific island ecosystems are notable for the high percentage of species found only in the region, and their biodiversity is both an important cultural resource for island people and a source of economic revenue through tourism. Terrestrial habitats and the goods and services they provide are threatened by rising temperatures, changes in rainfall, increased storminess, and land-use change. These changes promote the spread of invasive species and reduce the ability of habitats to support protected species and sustain human communities. Some species are expected to become extinct and others to decline to the point of requiring protection and costly management.

Key Message 3

Coastal Communities and Systems

The majority of Pacific island communities are confined to a narrow band of land within a few feet of sea level. Sea level rise is now beginning to threaten critical assets such as ecosystems, cultural sites and practices, economies, housing and energy, transportation, and other forms of infrastructure. By 2100, increases of 1–4 feet in global sea level are very likely, with even higher levels than the global average in the U.S.-Affiliated Pacific Islands. This would threaten the food and freshwater supply of Pacific island populations and jeopardize their continued sustainability and resilience. As sea level rise is projected to accelerate strongly after mid-century, adaptation strategies that are implemented sooner can better prepare communities and infrastructure for the most severe impacts.

Key Message 4

Oceans and Marine Resources

Fisheries, coral reefs, and the livelihoods they support are threatened by higher ocean temperatures and ocean acidification. Widespread coral reef bleaching and mortality have been occurring more frequently, and by mid-century these events are projected to occur annually, especially if current trends in emissions continue. Bleaching and acidification will result in loss of reef structure, leading to lower fisheries yields and loss of coastal protection and habitat. Declines in oceanic fishery productivity of up to 15% and 50% of current levels are projected by mid-century and 2100, respectively, under the higher scenario (RCP8.5).

Key Message 5

Indigenous Communities and Knowledge

Indigenous peoples of the Pacific are threatened by rising sea levels, diminishing freshwater availability, and shifting ecosystem services. These changes imperil communities' health, well-being, and modern livelihoods, as well as their familial relationships with lands, territories, and resources. Built on observations of climatic changes over time, the transmission and protection of traditional knowledge and practices, especially via the central role played by Indigenous women, are intergenerational, place-based, localized, and vital for ongoing adaptation and survival.

Key Message 6

Cumulative Impacts and Adaptation

Climate change impacts in the Pacific Islands are expected to amplify existing risks and lead to compounding economic, environmental, social, and cultural costs. In some locations, climate change impacts on ecological and social systems are projected to result in severe disruptions to livelihoods that increase the risk of human conflict or compel the need for migration. Early interventions, already occurring in some places across the region, can prevent costly and lengthy rebuilding of communities and livelihoods and minimize displacement and relocation.

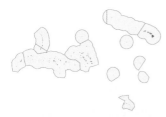

The U.S. Pacific Islands are culturally and environmentally diverse, treasured by the 1.9 million people who call them home. Pacific islands are particularly vulnerable to climate change impacts due to their exposure and isolation, small size, low elevation (in the case of atolls), and concentration of infrastructure and economy along the coasts.

A prevalent cause of year-to-year changes in climate patterns around the globe and in the Pacific Islands region is the El Niño–Southern Oscillation (ENSO). The El Niño and La Niña phases of ENSO can dramatically affect precipitation, air and ocean temperature, sea surface height, storminess, wave size, and trade winds. It is unknown exactly how the timing and intensity of ENSO will continue to change in the coming decades, but recent climate model results suggest a doubling in frequency of both El Niño and La Niña extremes in this century as compared to the 20th century under scenarios with more warming, including the higher scenario (RCP8.5).

On islands, all natural sources of freshwater come from rainfall received within their limited land areas. Severe droughts are common, making water shortage one of the most important climate-related risks in the region. As temperature continues to rise and cloud cover decreases in some areas, evaporation is expected to increase, causing both reduced water supply and higher water demand. Streamflow in Hawaiʻi has declined over approximately the past 100 years, consistent with observed decreases in rainfall.

The impacts of sea level rise in the Pacific include coastal erosion, episodic flooding, permanent inundation, heightened exposure to marine hazards, and saltwater intrusion to surface water and groundwater systems. Sea level rise will disproportionately affect the tropical Pacific and potentially exceed the global average.

Invasive species, landscape change, habitat alteration, and reduced resilience have resulted in extinctions and diminished ecosystem function. Inundation of atolls in the coming decades is projected to impact existing on-island ecosystems. Wildlife that relies on coastal habitats will likely also be severely impacted. In Hawaiʻi, coral reefs contribute an estimated $477 million to the local economy every year. Under projected warming of approximately 0.5°F per decade, all nearshore coral reefs in the Hawaiʻi and Pacific Islands region will experience annual bleaching before 2050. An ecosystem-based approach to international management of open ocean fisheries in the Pacific that incorporates climate-informed catch limits is expected to produce more realistic future harvest levels and enhance ecosystem resilience.

Indigenous communities of the Pacific derive their sense of identity from the islands. Emerging issues for Indigenous communities of the Pacific include the resilience of marine-managed areas and climate-induced human migration from their traditional lands. The rich body of traditional knowledge is place-based and localized and is useful in adaptation planning because it builds on intergenerational sharing of observations. Documenting the kinds of governance structures or decision-making hierarchies created for management of these lands and waters is also important as a learning tool that can be shared with other island communities.

Across the region, groups are coming together to minimize damage and disruption from coastal flooding and inundation as well as other

Climate Indicators and Impacts

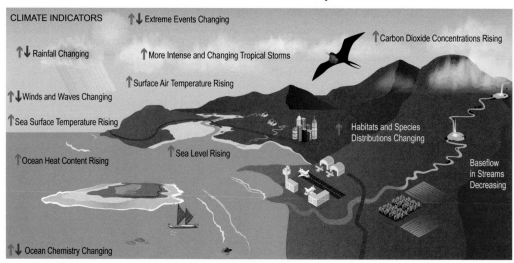

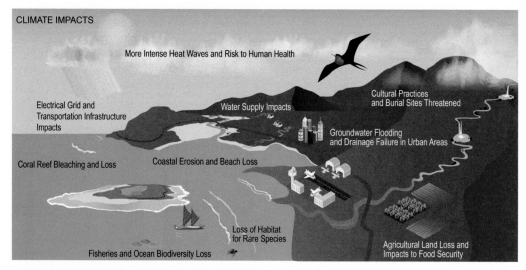

Monitoring regional indicator variables in the atmosphere, land, and ocean allows for tracking climate variability and change. (top) Observed changes in key climate indicators such as carbon dioxide concentration, sea surface temperatures, and species distributions in Hawai'i and the U.S.-Affiliated Pacific Islands result in (bottom) impacts to multiple sectors and communities, including built infrastructure, natural ecosystems, and human health. Connecting changes in climate indicators to how impacts are experienced is crucial in understanding and adapting to risks across different sectors. *From Figure 27.2 (Source: adapted from Keener et al. 2012).*

climate-related impacts. Social cohesion is already strong in many communities, making it possible to work together to take action. Early intervention can lower economic, environmental, social, and cultural costs and reduce or prevent conflict and displacement from ancestral land and resources.

For full chapter, including references and Traceable Accounts, see https://nca2018. globalchange.gov/chapter/hawaii-pacific.

Projected Onset of Annual Severe Coral Reef Bleaching

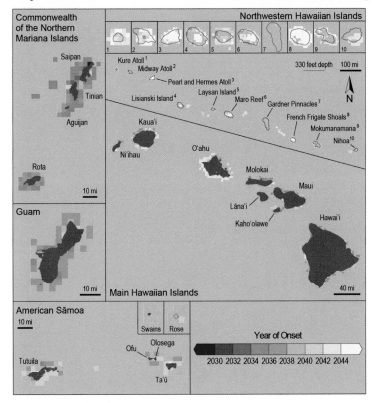

The figure shows the years when severe coral bleaching is projected to occur annually in the Hawai'i and U.S.-Affiliated Pacific Islands region under a higher scenario (RCP8.5). Darker colors indicate earlier projected onset of coral bleaching. Under projected warming of approximately 0.5°F per decade, all nearshore coral reefs in the region will experience annual bleaching before 2050. *From Figure 27.10 (Source: NOAA).*

Banner Photo Credits

18. Northeast: Bartram Bridge in Pennsylvania. © *Thomas James Caldwell/Flickr. CC BY-SA 2.0,* *https://creativecommons.org/licenses/by-sa/2.0/legalcode.*

19. Southeast: Red mangrove in Titusville, Florida. © *Katja Schulz/Flickr. CC BY 2.0,* *https://creativecommons.org/licenses/by/2.0/legalcode.*

20. U.S. Caribbean: San Juan, Puerto Rico. © *stevereidlphoto/ iStock/Getty Images*

21. Midwest: Carson, Wisconsin. © *William Garrett/Flickr. CC BY 2.0,* *https://creativecommons.org/licenses/by/2.0/legalcode.*

22. N. Great Plains: Cameron, Montana. *Paul Cross/U.S. Geological Survey.*

23. S. Great Plains: Whooping cranes in the Aransas National Wildlife Refuge in Texas. *Jon Noll/U.S. Department of Agriculture.*

24. Northwest: Four Lakes basin in White Cloud Peaks, Sawtooth National Forest, Idaho. *Mark Lisk/USDA Forest Service.*

25. Southwest: Low water levels in Lake Mead. © *Wayne Hsieh/ Flickr. CC BY-NC 2.0,* *https://creativecommons.org/licenses/by-nc/2.0/legalcode.*

26. Alaska: Anchorage, Alaska. © *Rocky Grimes/ istock/Getty Images.*

27. Hawai'i & Pacific Islands: Honolulu, Hawai'i. *NOAA Teacher at Sea Program, NOAA Ship Hi'ialakai.*

Note: Photos have been cropped from their original size in order to fit the report template.

Responses

Executive Summaries

Seawall surrounding Kivalina, Alaska

Key Message 1

Adaptation Implementation Is Increasing

Adaptation planning and implementation activities are occurring across the United States in the public, private, and nonprofit sectors. Since the Third National Climate Assessment, implementation has increased but is not yet commonplace.

Key Message 2

Climate Change Outpaces Adaptation Planning

Successful adaptation has been hindered by the assumption that climate conditions are and will be similar to those in the past. Incorporating information on current and future climate conditions into design guidelines, standards, policies, and practices would reduce risk and adverse impacts.

Key Message 3

Adaptation Entails Iterative Risk Management

Adaptation entails a continuing risk management process; it does not have an end point. With this approach, individuals and organizations of all types assess risks and vulnerabilities from climate and other drivers of change (such as economic, environmental, and societal), take actions to reduce those risks, and learn over time.

Key Message 4

Benefits of Proactive Adaptation Exceed Costs

Proactive adaptation initiatives—including changes to policies, business operations, capital investments, and other steps—yield benefits in excess of their costs in the near term, as well as over the long term. Evaluating adaptation strategies involves consideration of equity, justice, cultural heritage, the environment, health, and national security.

Key Message 5

New Approaches Can Further Reduce Risk

Integrating climate considerations into existing organizational and sectoral policies and practices provides adaptation benefits. Further reduction of the risks from climate change can be achieved by new approaches that create conditions for altering regulatory and policy environments, cultural and community resources, economic and financial systems, technology applications, and ecosystems.

Across the United States, many regions and sectors are already experiencing the direct effects of climate change. For these communities, climate impacts—from extreme storms made worse by sea level rise, to longer-lasting and more extreme heat waves, to increased numbers of wildfires and floods—are an immediate threat, not a far-off possibility. Because these impacts are expected to increase over time, communities throughout the United States face the challenge not only of reducing greenhouse gas emissions, but also of adapting to current and future climate change to help mitigate climate risks.

Adaptation takes place at many levels—national and regional but mainly local—as governments, businesses, communities, and individuals respond to today's altered climate conditions and prepare for future change based on the specific climate impacts relevant to their geography and vulnerability. Adaptation has five general stages: awareness, assessment, planning, implementation, and monitoring and evaluation. These phases naturally build on one another, though they

are often not executed sequentially and the terminology may vary. The Third National Climate Assessment (released in 2014) found the first three phases underway throughout the United States but limited in terms of on-the-ground implementation. Since then, the scale and scope of adaptation implementation have increased, but in general, adaptation implementation is not yet commonplace.

One important aspect of adaptation is the ability to anticipate future climate impacts and plan accordingly. Public- and private-sector decision-makers have traditionally made plans assuming that the current and future climate in their location will resemble that of the recent past. This assumption is no longer reliably true. Increasingly, planners, builders, engineers, architects, contractors, developers, and other individuals are recognizing the need to take current and projected climate conditions into account in their decisions about the location and design of buildings and infrastructure, engineering standards, insurance rates, property values, land-use plans, disaster response preparations, supply chains, and cropland and forest management.

In anticipating and planning for climate change, decision-makers practice a form of risk assessment known as iterative risk management. Iterative risk management emphasizes that the process of anticipating and responding to climate change does not constitute a single set of judgments at any point in time; rather, it is an ongoing cycle of assessment, action, reassessment, learning, and response. In the adaptation context, public- and private-sector actors manage climate risk using three types of actions: reducing exposure, reducing sensitivity, and increasing adaptive capacity.

Climate risk management includes some attributes and tactics that are familiar to most businesses and local governments, since these organizations already commonly manage or design for a variety of weather-related risks, including coastal and inland storms, heat waves, water availability threats, droughts, and floods. However, successful adaptation also requires the often unfamiliar challenge of using information on current and future climate, rather than past climate, which can prove difficult for those lacking experience with climate change datasets and concepts. In addition, many professional practices and guidelines, as well as legal requirements, still call for the use of data based on past climate. Finally, factors such as access to resources, culture, governance, and available information can affect not only the risk faced by different populations but also the best ways to reduce their risks.

Achieving the benefits of adaptation can require up-front investments to achieve longer-term savings, engaging with differing stakeholder interests and values, and planning in the face of uncertainty. But adaptation also presents challenges, including difficulties in obtaining the necessary funds, insufficient information and relevant expertise, and jurisdictional mismatches.

In general, adaptation can generate significant benefits in excess of its costs. Benefit–cost analysis can help guide organizations toward actions that most efficiently reduce risks, in particular those that, if not addressed, could prove extremely costly in the future. Beyond those attributes explicitly measured by benefit–cost analysis, effective adaptation can also enhance social welfare in many ways that can be difficult to quantify and that people will value differently, including improving economic opportunity, health, equity, security, education, social connectivity, and sense of place, as well as safeguarding cultural resources and practices and environmental quality.

A significant portion of climate risk can be addressed by mainstreaming; that is, integrating climate adaptation into existing organizational and sectoral investments, policies, and practices, such as planning, budgeting, policy development, and operations and maintenance. Mainstreaming of climate adaptation into existing decision processes has already begun in many areas, such as financial risk reporting, capital investment planning, engineering standards, military planning, and disaster risk management. Further reduction of the risks from climate change, in particular those that arise from futures with high levels of greenhouse gas emissions, calls for new approaches that create conditions for altering regulatory and policy environments, cultural and community resources, economic and financial systems, technology applications, and ecosystems.

For full chapter, including references and Traceable Accounts, see https://nca2018. globalchange.gov/chapter/adaptation.

Five Adaptation Stages and Progress

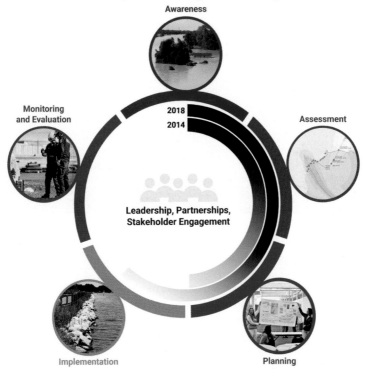

The figure illustrates the adaptation iterative risk management process. The gray arced lines compare the current status of implementing this process with the status reported by the Third National Climate Assessment in 2014. Darker color indicates more activity. *From Figure 28.1 (Source: adapted from National Research Council, 2010. Used with permission from the National Academies Press, © 2010, National Academy of Sciences. Image credits, clockwise from top: National Weather Service; USGS; Armando Rodriguez, Miami-Dade County; Dr. Neil Berg, MARISA; Bill Ingalls, NASA).*

Jasper, New York

Key Message 1

Mitigation-Related Activities Within the United States

Mitigation-related activities are taking place across the United States at the federal, state, and local levels as well as in the private sector. Since the Third National Climate Assessment, a growing number of states, cities, and businesses have pursued or deepened initiatives aimed at reducing emissions.

Key Message 2

The Risks of Inaction

In the absence of more significant global mitigation efforts, climate change is projected to impose substantial damages on the U.S. economy, human health, and the environment. Under scenarios with high emissions and limited or no adaptation, annual losses in some sectors are estimated to grow to hundreds of billions of dollars by the end of the century. It is very likely that some physical and ecological impacts will be irreversible for thousands of years, while others will be permanent.

Key Message 3

Avoided or Reduced Impacts Due to Mitigation

Many climate change impacts and associated economic damages in the United States can be substantially reduced over the course of the 21st century through global-scale reductions in greenhouse gas emissions, though the magnitude and timing of avoided risks vary by sector and region. The effect of near-term emissions mitigation on reducing risks is expected to become apparent by mid-century and grow substantially thereafter.

Key Message 4

Interactions Between Mitigation and Adaptation

Interactions between mitigation and adaptation are complex and can lead to benefits, but they also have the potential for adverse consequences. Adaptation can complement mitigation to substantially reduce exposure and vulnerability to climate change in some sectors. This complementarity is especially important given that a certain degree of climate change due to past and present emissions is unavoidable.

Current and future emissions of greenhouse gases, and thus emission mitigation actions, are crucial for determining future risks and impacts of climate change to society. The scale of risks that can be avoided through mitigation actions is influenced by the magnitude of emissions reductions, the timing of those reductions, and the relative mix of mitigation strategies for emissions of long-lived greenhouse gases (namely, carbon dioxide), short-lived greenhouse gases (such as methane), and land-based biologic carbon. Many actions at national, regional, and local scales are underway to reduce greenhouse gas emissions, including efforts in the private sector.

Climate change is projected to significantly damage human health, the economy, and the environment in the United States, particularly under a future with high greenhouse gas emissions. A collection of frontier research initiatives is underway to improve understanding and quantification of climate impacts. These studies have been designed across a variety of sectoral and spatial scales and feature the use of internally consistent climate and socioeconomic scenarios. Recent findings from these multisector modeling frameworks demonstrate substantial and far-reaching changes over the course of the 21st century—and particularly at the end of the century—with negative consequences for a large majority of sectors, including infrastructure and human health. For sectors where positive effects are observed in some regions or for specific time periods,

the effects are typically dwarfed by changes happening overall within the sector or at broader scales.

Recent studies also show that many climate change impacts in the United States can be substantially reduced over the course of the 21st century through global-scale reductions in greenhouse gas emissions. While the difference in climate outcomes between scenarios is more modest through the first half of the century, the effect of mitigation in avoiding climate change impacts typically becomes clear by 2050 and increases substantially in magnitude thereafter. Research supports that early and substantial mitigation offers a greater chance of avoiding increasingly adverse impacts.

The reduction of climate change risk due to mitigation also depends on assumptions about how adaptation changes the exposure and vulnerability of the population. Physical damages to coastal property and transportation infrastructure are particularly sensitive to adaptation assumptions, with proactive measures estimated to be capable of reducing damages by large fractions. Because society is already committed to a certain amount of future climate change due to past and present emissions and because mitigation activities cannot avoid all climate-related risks, mitigation and adaptation activities can be considered complementary strategies. However, adaptation can require large up-front costs and long-term commitments for maintenance, and uncertainty exists in some sectors regarding the applicability and effectiveness of adaptation in reducing risk. Interactions between adaptation and mitigation strategies can result in benefits or adverse consequences. While uncertainties still remain, advancements in the modeling of climate and economic impacts, including current understanding of adaptation pathways, are increasingly providing new capabilities to understand and quantify future effects.

For full chapter, including references and Traceable Accounts, see https://nca2018. globalchange.gov/chapter/mitigation.

Projected Damages and Potential for Risk Reduction by Sector

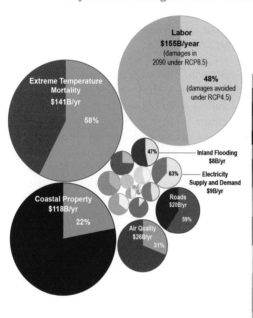

Annual Economic Damages in 2090		
Sector	Annual damages under RCP8.5	Damages avoided under RCP4.5
Labor	$155B	48%
Extreme Temperature Mortality◊	$141B	58%
Coastal Property◊	$118B	22%
Air Quality	$26B	31%
Roads◊	$20B	59%
Electricity Supply and Demand	$9B	63%
Inland Flooding	$8B	47%
Urban Drainage	$6B	26%
Rail◊	$6B	36%
Water Quality	$5B	35%
Coral Reefs	$4B	12%
West Nile Virus	$3B	47%
Freshwater Fish	$3B	44%
Winter Recreation	$2B	107%
Bridges	$1B	48%
Munic. and Industrial Water Supply	$316M	33%
Harmful Algal Blooms	$199M	45%
Alaska Infrastructure◊	$174M	53%
Shellfish*	$23M	57%
Agriculture*	$12M	11%
Aeroallergens*	$1M	57%
Wildfire	-$106M	-134%

The total area of each circle represents the projected annual economic damages (in 2015 dollars) under a higher scenario (RCP8.5) in 2090 relative to a no-change scenario. The decrease in damages under a lower scenario (RCP4.5) compared to RCP8.5 is shown in the lighter-shaded area of each circle. Where applicable, sectoral results assume population change over time, which in the case of winter recreation leads to positive effects under RCP4.5, as increased visitors outweigh climate losses. Importantly, many sectoral damages from climate change are not included here, and many of the reported results represent only partial valuations of the total physical damages. See EPA 2017 for ranges surrounding the central estimates presented in the figure; results assume limited or no adaptation. Adaptation was shown to reduce overall damages in sectors identified with the diamond symbol but was not directly modeled in, or relevant to, all sectors. Asterisks denote sectors with annual damages that may not be visible at the given scale. Only one impact (wildfire) shows very small positive effects, owing to projected landscape-scale shifts to vegetation with longer fire return intervals (see Ch. 6: Forests for a discussion on the weight of evidence regarding projections of future wildfire activity). The online version of this figure includes value ranges for numbers in the table. Due to space constraints, the ranges are not included here. *From Figure 29.2 (Source: adapted from EPA 2017).*

Banner Photo Credits

28. Adaptation: Seawall surrounding Kivalina, Alaska. © *ShoreZone. CC BY 3.0, https://creativecommons.org/ licenses/by/3.0/legalcode*.

29. Mitigation: Jasper, New York. © *John Getchel/Flickr. CC BY-NC 2.0, https://creativecommons.org/licenses/ by-nc/2.0/legalcode*.

Note: Photos have been cropped from their original size in order to fit the report template.

Fourth National Climate Assessment Author Teams

1. Overview

Federal Coordinating Lead Author
David Reidmiller, U.S. Global Change Research Program

Chapter Lead
Alexa Jay, U.S. Global Change Research Program

Chapter Authors
Christopher W. Avery, U.S. Global Change Research Program
Daniel Barrie, National Oceanic and Atmospheric Administration
Apurva Dave, U.S. Global Change Research Program
Benjamin DeAngelo, National Oceanic and Atmospheric Administration
Matthew Dzaugis, U.S. Global Change Research Program
Michael Kolian, U.S. Environmental Protection Agency
Kristin Lewis, U.S. Global Change Research Program
Katie Reeves, U.S. Global Change Research Program
Darrell Winner, U.S. Environmental Protection Agency

2. Our Changing Climate

Federal Coordinating Lead Authors
David R. Easterling, NOAA National Centers for Environmental Information
David W. Fahey, NOAA Earth System Research Laboratory

Chapter Lead
Katharine Hayhoe, Texas Tech University

Chapter Authors
Sarah Doherty, University of Washington
James P. Kossin, NOAA National Centers for Environmental Information
William V. Sweet, NOAA National Ocean Service
Russell S. Vose, NOAA National Centers for Environmental Information
Michael F. Wehner, Lawrence Berkeley National Laboratory
Donald J. Wuebbles, University of Illinois

Technical Contributors
Robert E. Kopp, Rutgers University
Kenneth E. Kunkel, North Carolina State University
John Nielsen-Gammon, Texas A&M University

Review Editor
Linda O. Mearns, National Center for Atmospheric Research

USGCRP Coordinators
David J. Dokken, Senior Program Officer
David Reidmiller, Director

3. Water

Federal Coordinating Lead Authors
Thomas Johnson, U.S. Environmental Protection Agency
Peter Colohan, National Oceanic and Atmospheric Administration

Chapter Lead
Upmanu Lall, Columbia University

Chapter Authors
Amir AghaKouchak, University of California, Irvine
Sankar Arumugam, North Carolina State University
Casey Brown, University of Massachusetts
Gregory McCabe, U.S. Geological Survey
Roger Pulwarty, National Oceanic and Atmospheric Administration

Review Editor
Minxue He, California Department of Water Resources

USGCRP Coordinators
Kristin Lewis, Senior Scientist
Allyza Lustig, Program Coordinator

4. Energy Supply, Delivery, and Demand

Federal Coordinating Lead Author
Craig D. Zamuda, U.S. Department of Energy, Office of Policy

Chapter Lead
Craig D. Zamuda, U.S. Department of Energy, Office of Policy

Chapter Authors
Daniel E. Bilello, National Renewable Energy Laboratory
Guenter Conzelmann, Argonne National Laboratory
Ellen Mecray, National Oceanic and Atmospheric Administration
Ann Satsangi, U.S. Department of Energy, Office of Fossil Energy
Vincent Tidwell, Sandia National Laboratories
Brian J. Walker, U.S. Department of Energy, Office of Energy Efficiency and Renewable Energy

Review Editor
Sara C. Pryor, Cornell University

USGCRP Coordinators
Natalie Bennett, Adaptation and Assessment Analyst
Christopher W. Avery, Senior Manager

5. Land Cover and Land-Use Change

Federal Coordinating Lead Author
Thomas Loveland, U.S. Geological Survey

Chapter Lead
Benjamin M. Sleeter, U.S. Geological Survey

Chapter Authors
James Wickham, U.S. Environmental Protection Agency
Grant Domke, U.S. Forest Service
Nate Herold, National Oceanic and Atmospheric
    Administration
Nathan Wood, U.S. Geological Survey

Technical Contributors
Tamara S. Wilson, U.S. GeologicWal Survey
Jason Sherba, U.S. Geological Survey

Review Editor
Georgine Yorgey, Washington State University

USGCRP Coordinators
Susan Aragon-Long, Senior Scientist
Christopher W. Avery, Senior Manager

6. Forests

Federal Coordinating Lead Authors
James M. Vose, U.S. Forest Service, Southern
    Research Station
David L. Peterson, U.S. Forest Service, Pacific Northwest
    Research Station

Chapter Leads
James M. Vose, U.S. Forest Service, Southern
    Research Station
David L. Peterson, U.S. Forest Service, Pacific Northwest
    Research Station

Chapter Authors
Grant M. Domke, U.S. Forest Service, Northern
    Research Station
Christopher J. Fettig, U.S. Forest Service, Pacific
    Southwest Research Station
Linda A. Joyce, U.S. Forest Service, Rocky Mountain
    Research Station
Robert E. Keane, U.S. Forest Service, Rocky Mountain
    Research Station
Charles H. Luce, U.S. Forest Service, Rocky Mountain
    Research Station
Jeffrey P. Prestemon, U.S. Forest Service, Southern
    Research Station

Technical Contributors
Lawrence E. Band, University of Virginia
James S. Clark, Duke University
Nicolette E. Cooley, Northern Arizona University
Anthony D'Amato, University of Vermont
Jessica E. Halofsky, University of Washington

Review Editor
Gregg Marland, Appalachian State University

USGCRP Coordinators
Natalie Bennett, Adaptation and Assessment Analyst
Susan Aragon-Long, Senior Scientist

7. Ecosystems, Ecosystem Services, and Biodiversity

Federal Coordinating Lead Authors
Shawn Carter, U.S. Geological Survey
Jay Peterson, National Oceanic and Atmospheric
    Administration

Chapter Leads
Douglas Lipton, National Oceanic and Atmospheric
    Administration
Madeleine A. Rubenstein. U.S. Geological Survey
Sarah R. Weiskopf, U.S. Geological Survey

Chapter Authors
Lisa Crozier, National Oceanic and Atmospheric
    Administration
Michael Fogarty, National Oceanic and Atmospheric
    Administration
Sarah Gaichas, National Oceanic and Atmospheric
    Administration
Kimberly J. W. Hyde, National Oceanic and Atmospheric
    Administration
Toni Lyn Morelli, U.S. Geological Survey
Jeffrey Morisette, U.S. Department of the Interior,
    National Invasive Species Council Secretariat
Hassan Moustahfid, National Oceanic and Atmospheric
    Administration
Roldan Muñoz, National Oceanic and Atmospheric
    Administration
Rajendra Poudel, National Oceanic and Atmospheric
    Administration
Michelle D. Staudinger, U.S. Geological Survey
Charles Stock, National Oceanic and Atmospheric
    Administration
Laura Thompson, U.S. Geological Survey
Robin Waples, National Oceanic and Atmospheric
    Administration
Jake F. Weltzin, U.S. Geological Survey

Review Editor
Gregg Marland, Appalachian State University

USGCRP Coordinators
Matthew Dzaugis, Program Coordinator
Allyza Lustig, Program Coordinator

8. Coastal Effects

Federal Coordinating Lead Authors
Jeffrey Payne, National Oceanic and Atmospheric
Administration
William V. Sweet, National Oceanic and Atmospheric
Administration

Chapter Lead
Elizabeth Fleming, U.S. Army Corps of Engineers

Chapter Authors
Michael Craghan, U.S. Environmental Protection Agency
John Haines, U.S. Geological Survey
Juliette Finzi Hart, U.S. Geological Survey
Heidi Stiller, National Oceanic and Atmospheric
Administration
Ariana Sutton-Grier, National Oceanic and Atmospheric
Administration

Review Editor
Michael Kruk, ERT, Inc.

USGCRP Coordinators
Matthew Dzaugis, Program Coordinator
Christopher W. Avery, Senior Manager
Allyza Lustig, Program Coordinator
Fredric Lipschultz, Senior Scientist and Regional
Coordinator

9. Oceans and Marine Resources

Federal Coordinating Lead Authors
Roger B. Griffis, National Oceanic and Atmospheric
Administration
Elizabeth B. Jewett, National Oceanic and Atmospheric
Administration

Chapter Lead
Andrew J. Pershing, Gulf of Maine Research Institute

Chapter Authors
C. Taylor Armstrong, National Oceanic and Atmospheric
Administration
John F. Bruno, University of North Carolina at Chapel Hill
D. Shallin Busch, National Oceanic and Atmospheric
Administration
Alan C. Haynie, National Oceanic and Atmospheric
Administration
Samantha A. Siedlecki, University of Washington (now
University of Connecticut)
Desiree Tommasi, University of California, Santa Cruz

Technical Contributor
Vicky W. Y. Lam, University of British Columbia

Review Editor
Sarah R. Cooley, Ocean Conservancy

USGCRP Coordinators
Fredric Lipschultz, Senior Scientist and Regional
Coordinator
Apurva Dave, International Coordinator and
Senior Analyst

10. Agriculture and Rural Communities

Federal Coordinating Lead Author
Carolyn Olson, U.S. Department of Agriculture

Chapter Leads
Prasanna Gowda, USDA Agricultural Research Service
Jean L. Steiner, USDA Agricultural Research Service

Chapter Authors
Tracey Farrigan, USDA Economic Research Service
Michael A. Grusak, USDA Agricultural Research Service
Mark Boggess, USDA Agricultural Research Service

Review Editor
Georgine Yorgey, Washington State University

USGCRP Coordinators
Susan Aragon-Long, Senior Scientist
Allyza Lustig, Program Coordinator

11. Built Environment, Urban Systems, and Cities

Federal Coordinating Lead Author
Susan Julius, U.S. Environmental Protection Agency

Chapter Lead
Keely Maxwell, U.S. Environmental Protection Agency

Chapter Authors
Anne Grambsch, U.S. Environmental Protection
Agency (Retired)
Ann Kosmal, U.S. General Services Administration
Libby Larson, National Aeronautics and
Space Administration
Nancy Sonti, U.S. Forest Service

Technical Contributors
Julie Blue, Eastern Research Group, Inc.
Kevin Bush, U.S. Department of Housing and Urban
Development (through August 2017)

Review Editor
Jesse Keenan, Harvard University

USGCRP Coordinators
Natalie Bennett, Adaptation and Assessment Analyst
Fredric Lipschultz, Senior Scientist and Regional
Coordinator

12. Transportation

Federal Coordinating Lead Author
Michael Culp, U.S. Department of Transportation, Federal Highway Administration

Chapter Lead
Jennifer M. Jacobs, University of New Hampshire

Chapter Authors
Lia Cattaneo, Harvard University (formerly U.S. Department of Transportation)
Paul Chinowsky, University of Colorado Boulder
Anne Choate, ICF
Susanne DesRoches, New York City Mayor's Office of Recovery and Resiliency and Office of Sustainability
Scott Douglass, South Coast Engineers
Rawlings Miller, WSP (formerly U.S. Department of Transportation Volpe Center)

Review Editor
Jesse Keenan, Harvard University

USGCRP Coordinators
Allyza Lustig, Program Coordinator
Kristin Lewis, Senior Scientist

13. Air Quality

Federal Coordinating Lead Author
Christopher G. Nolte, U.S. Environmental Protection Agency

Chapter Lead
Christopher G. Nolte, U.S. Environmental Protection Agency

Chapter Authors
Patrick D. Dolwick, U.S. Environmental Protection Agency
Neal Fann, U.S. Environmental Protection Agency
Larry W. Horowitz, National Oceanic and Atmospheric Administration
Vaishali Naik, National Oceanic and Atmospheric Administration
Robert W. Pinder, U.S. Environmental Protection Agency
Tanya L. Spero, U.S. Environmental Protection Agency
Darrell A. Winner, U.S. Environmental Protection Agency
Lewis H. Ziska, U.S. Department of Agriculture

Review Editor
David D'Onofrio, Atlanta Regional Commission

USGCRP Coordinators
Ashley Bieniek-Tobasco, Health Program Coordinator
Sarah Zerbonne, Adaptation and Decision Science Coordinator
Christopher W. Avery, Senior Manager

14. Human Health

Federal Coordinating Lead Authors
John M. Balbus, National Institute of Environmental Health Sciences
George Luber, Centers for Disease Control and Prevention

Chapter Lead
Kristie L. Ebi, University of Washington

Chapter Authors
Aparna Bole, University Hospitals Rainbow Babies & Children's Hospital, Ohio
Allison Crimmins, U.S. Environmental Protection Agency
Gregory Glass, University of Florida
Shubhayu Saha, Centers for Disease Control and Prevention
Mark M. Shimamoto, American Geophysical Union
Juli Trtanj, National Oceanic and Atmospheric Administration
Jalonne L. White-Newsome, The Kresge Foundation

Technical Contributors
Stasia Widerynski, Centers for Disease Control and Prevention

Review Editor
David D'Onofrio, Atlanta Regional Commission

USGCRP Coordinators
Ashley Bieniek-Tobasco, Health Program Coordinator
Sarah Zerbonne, Adaptation and Decision Science Coordinator
Natalie Bennett, Adaptation and Assessment Analyst
Christopher W. Avery, Senior Manager

15. Tribes and Indigenous Peoples

Federal Coordinating Lead Author
Rachael Novak, U.S. Department of the Interior, Bureau of Indian Affairs

Chapter Lead
Lesley Jantarasami, Oregon Department of Energy

Chapter Authors
Roberto Delgado, National Institutes of Health
Elizabeth Marino, Oregon State University–Cascades
Shannon McNeeley, North Central Climate Adaptation Science Center and Colorado State University
Chris Narducci, U.S. Department of Housing and Urban Development
Julie Raymond-Yakoubian, Kawerak, Inc.
Loretta Singletary, University of Nevada, Reno
Kyle Powys Whyte, Michigan State University

Review Editor
Karen Cozzetto, Northern Arizona University

USGCRP Coordinators
Susan Aragon-Long, Senior Scientist
Allyza Lustig, Program Coordinator

16. Climate Effects on U.S. International Interests

Federal Coordinating Lead Author
Meredith Muth, National Oceanic and Atmospheric Administration

Chapter Lead
Joel B. Smith, Abt Associates

Chapter Authors
Alice Alpert, U.S. Department of State
James L. Buizer, University of Arizona
Jonathan Cook, World Resources Institute (formerly U.S. Agency for International Development)
Apurva Dave, U.S. Global Change Research Program/ICF
John Furlow, International Research Institute for Climate and Society, Columbia University
Kurt Preston, U.S. Department of Defense
Peter Schultz, ICF
Lisa Vaughan, National Oceanic and Atmospheric Administration

Review Editor
Diana Liverman, University of Arizona

USGCRP Coordinators
Apurva Dave, International Coordinator and Senior Analyst

17. Sector Interactions, Multiple Stressors, and Complex Systems

Federal Coordinating Lead Authors
Leah Nichols, National Science Foundation
Robert Vallario, U.S. Department of Energy

Chapter Lead
Leon Clarke, Pacific Northwest National Laboratory

Chapter Authors
Mohamad Hejazi, Pacific Northwest National Laboratory
Jill Horing, Pacific Northwest National Laboratory
Anthony C. Janetos, Boston University
Katharine Mach, Stanford University
Michael Mastrandrea, Carnegie Institution for Science
Marilee Orr, U.S. Department of Homeland Security
Benjamin L. Preston, Rand Corporation
Patrick Reed, Cornell University
Ronald D. Sands, U.S. Department of Agriculture
Dave D. White, Arizona State University

Review Editor
Kai Lee, Williams College (Emeritus) and the Packard Foundation (Retired)

USGCRP Coordinators
Kristin Lewis, Senior Scientist
Natalie Bennett, Adaptation and Assessment Analyst

18. Northeast

Federal Coordinating Lead Author
Ellen L. Mecray, National Oceanic and Atmospheric Administration

Chapter Lead
Lesley-Ann L. Dupigny-Giroux, University of Vermont

Chapter Authors
Mary D. Lemcke-Stampone, University of New Hampshire
Glenn A. Hodgkins, U.S. Geological Survey
Erika E. Lentz, U.S. Geological Survey
Katherine E. Mills, Gulf of Maine Research Institute
Erin D. Lane, U.S. Department of Agriculture
Rawlings Miller, WSP (formerly U.S. Department of Transportation Volpe Center)
David Y. Hollinger, U.S. Department of Agriculture
William D. Solecki, City University of New York–Hunter College
Gregory A. Wellenius, Brown University
Perry E. Sheffield, Icahn School of Medicine at Mount Sinai
Anthony B. MacDonald, Monmouth University
Christopher Caldwell, College of Menominee Nation

Technical Contributors
Zoe P. Johnson, U.S. Department of Defense, Naval Facilities Engineering Command (formerly NOAA Chesapeake Bay Office)
Amanda Babson, U.S. National Park Service
Elizabeth Pendleton, U.S. Geological Survey
Benjamin T. Gutierrez, U.S. Geological Survey
Joseph Salisbury, University of New Hampshire
Andrew Sven McCall Jr., University of Vermont
E. Robert Thieler, U.S. Geological Survey
Sara L. Zeigler, U.S. Geological Survey

Review Editor
Jayne F. Knott, University of New Hampshire

USGCRP Coordinators
Christopher W. Avery, Senior Manager
Matthew Dzaugis, Program Coordinator
Allyza Lustig, Program Coordinator

19. Southeast

Federal Coordinating Lead Author
Adam Terando, U.S. Geological Survey, Southeast Climate Adaptation Science Center

Chapter Lead
Lynne Carter, Louisiana State University

Chapter Authors
Kirstin Dow, University of South Carolina
Kevin Hiers, Tall Timbers Research Station
Kenneth E. Kunkel, North Carolina State University
Aranzazu Lascurain, North Carolina State University
Doug Marcy, National Oceanic and Atmospheric Administration
Michael Osland, U.S. Geological Survey
Paul Schramm, Centers for Disease Control and Prevention

Technical Contributors
Vincent Brown, Louisiana State University
Barry Keim, Louisiana State University
Julie K. Maldonado, Livelihoods Knowledge Exchange Network
Colin Polsky, Florida Atlantic University
April Taylor, Chickasaw Nation

Review Editor
Alessandra Jerolleman, Jacksonville State University

USGCRP Coordinators
Allyza Lustig, Program Coordinator
Matthew Dzaugis, Program Coordinator
Natalie Bennett, Adaptation and Assessment Analyst

20. U.S. Caribbean

Federal Coordinating Lead Author
William A. Gould, USDA Forest Service International Institute of Tropical Forestry

Chapter Lead
Ernesto L. Díaz, Department of Natural and Environmental Resources, Coastal Zone Management Program

Chapter Authors
Nora L. Álvarez-Berríos, USDA Forest Service International Institute of Tropical Forestry
Felix Aponte-González, Aponte, Aponte & Asociados
Wayne Archibald, Archibald Energy Group
Jared Heath Bowden, Department of Applied Ecology, North Carolina State University
Lisamarie Carrubba, NOAA Fisheries, Office of Protected Resources
Wanda Crespo, Estudios Técnicos, Inc.
Stephen Joshua Fain, USDA Forest Service International Institute of Tropical Forestry
Grizelle González, USDA Forest Service International Institute of Tropical Forestry
Annmarie Goulbourne, Environmental Solutions Limited
Eric Harmsen, Department of Agricultural and Biosystems Engineering, University of Puerto Rico
Azad Henareh Khalyani, Natural Resource Ecology Laboratory, Colorado State University
Eva Holupchinski, USDA Forest Service International Institute of Tropical Forestry

James P. Kossin, National Oceanic and Atmospheric Administration
Amanda J. Leinberger, Center for Climate Adaptation Science and Solutions, University of Arizona
Vanessa I. Marrero-Santiago, Department of Natural and Environmental Resources, Coastal Zone Management Program
Odalys Martínez-Sánchez, NOAA National Weather Service
Kathleen McGinley, USDA Forest Service International Institute of Tropical Forestry
Melissa Meléndez Oyola, University of New Hampshire
Pablo Méndez-Lázaro, University of Puerto Rico
Julio Morell, University of Puerto Rico
Isabel K. Parés-Ramos, USDA Forest Service International Institute of Tropical Forestry
Roger Pulwarty, National Oceanic and Atmospheric Administration
William V. Sweet, NOAA National Ocean Service
Adam Terando, U.S. Geological Survey, Southeast Climate Adaptation Science Center
Sigfredo Torres-González, U.S. Geological Survey (Retired)

Technical Contributors
Mariano Argüelles, Puerto Rico Department of Agriculture
Gabriela Bernal-Vega, University of Puerto Rico
Roberto Moyano, Estudios Técnicos, Inc.
Pedro Nieves, USVI Coastal Zone Management
Aurelio Mercado-Irizarry, University of Puerto Rico
Dominique Davíd-Chavez, Colorado State University

Review Editor
Jess K. Zimmerman, University of Puerto Rico

USGCRP Coordinators
Allyza Lustig, Program Coordinator
Apurva Dave, International Coordinator and Senior Analyst
Christopher W. Avery, Senior Manager

21. Midwest

Federal Coordinating Lead Author
Chris Swanston, USDA Forest Service

Chapter Lead
Jim Angel, Prairie Research Institute, University of Illinois

Chapter Authors
Barbara Mayes Boustead, National Oceanic and Atmospheric Administration
Kathryn C. Conlon, Centers for Disease Control and Prevention
Kimberly R. Hall, The Nature Conservancy
Jenna L. Jorns, University of Michigan, Great Lakes Integrated Sciences and Assessments
Kenneth E. Kunkel, North Carolina State University

Maria Carmen Lemos, University of Michigan, Great Lakes Integrated Sciences and Assessments
Brent Lofgren, National Oceanic and Atmospheric Administration
Todd A. Ontl, USDA Forest Service, Northern Forests Climate Hub
John Posey, East West Gateway Council of Governments
Kim Stone, Great Lakes Indian Fish and Wildlife Commission (through January 2018)
Eugene Takle, Iowa State University
Dennis Todey, USDA, Midwest Climate Hub

Technical Contributors
Katherine Browne, University of Michigan
Melonee Montano, Great Lakes Indian Fish and Wildlife Commission
Hannah Panci, Great Lakes Indian Fish and Wildlife Commission
Jason Vargo, University of Wisconsin
Madeline R. Magee, University of Wisconsin–Madison

Review Editor
Thomas Bonnot, University of Missouri

USGCRP Coordinators
Kristin Lewis, Senior Scientist
Allyza Lustig, Program Coordinator
Katie Reeves, Engagement and Communications Lead

22. Northern Great Plains

Federal Coordinating Lead Author
Doug Kluck, National Oceanic and Atmospheric Administration

Chapter Lead
Richard T. Conant, Colorado State University

Chapter Authors
Mark Anderson, U.S. Geological Survey
Andrew Badger, University of Colorado
Barbara Mayes Boustead, National Oceanic and Atmospheric Administration
Justin Derner, U.S. Department of Agriculture
Laura Farris, U.S. Environmental Protection Agency
Michael Hayes, University of Nebraska
Ben Livneh, University of Colorado
Shannon McNeeley, North Central Climate Adaptation Science Center and Colorado State University
Dannele Peck, U.S. Department of Agriculture
Martha Shulski, University of Nebraska
Valerie Small, University of Arizona

Review Editor
Kirsten de Beurs, University of Oklahoma

USGCRP Coordinators
Allyza Lustig, Program Coordinator
Kristin Lewis, Senior Scientist

23. Southern Great Plains

Federal Coordinating Lead Author
Bill Bartush, U.S. Fish and Wildlife Service

Chapter Lead
Kevin Kloesel, University of Oklahoma

Chapter Authors
Jay Banner, University of Texas at Austin
David Brown, USDA-ARS Grazinglands Research Laboratory
Jay Lemery, University of Colorado
Xiaomao Lin, Kansas State University
Cindy Loeffler, Texas Parks and Wildlife Department
Gary McManus, Oklahoma Climatological Survey
Esther Mullens, DOI South Central Climate Adaptation Science Center
John Nielsen-Gammon, Texas A&M University
Mark Shafer, NOAA-RISA Southern Climate Impacts Planning Program
Cecilia Sorensen, University of Colorado
Sid Sperry, Oklahoma Association of Electric Cooperatives
Daniel Wildcat, Haskell Indian Nations University
Jadwiga Ziolkowska, University of Oklahoma

Technical Contributor
Katharine Hayhoe, Texas Tech University

Review Editor
Ellu Nasser, Adaptation International

USGCRP Coordinators
Susan Aragon-Long, Senior Scientist
Christopher W. Avery, Senior Manager

24. Northwest

Federal Coordinating Lead Author
Charles Luce, USDA Forest Service

Chapter Lead
Christine May, Silvestrum Climate Associates

Chapter Authors
Joe Casola, Climate Impacts Group, University of Washington
Michael Chang, Makah Tribe
Jennifer Cuhaciyan, Bureau of Reclamation
Meghan Dalton, Oregon State University
Scott Lowe, Boise State University
Gary Morishima, Quinault Indian Nation
Philip Mote, Oregon State University
Alexander (Sascha) Petersen, Adaptation International
Gabrielle Roesch-McNally, USDA Forest Service
Emily York, Oregon Health Authority

Review Editor
Beatrice Van Horne, USDA Forest Service, Northwest Climate Hub

USGCRP Coordinators
Natalie Bennett, Adaptation and Assessment Analyst
Christopher W. Avery, Senior Manager
Susan Aragon-Long, Senior Scientist

25. Southwest

Federal Coordinating Lead Author
Patrick Gonzalez, U.S. National Park Service

Chapter Lead
Gregg M. Garfin, University of Arizona

Chapter Authors
David D. Breshears, University of Arizona
Keely M. Brooks, Southern Nevada Water Authority
Heidi E. Brown, University of Arizona
Emile H. Elias, U.S. Department of Agriculture
Amrith Gunasekara, California Department of Food and Agriculture
Nancy Huntly, Utah State University
Julie K. Maldonado, Livelihoods Knowledge Exchange Network
Nathan J. Mantua, National Oceanic and Atmospheric Administration
Helene G. Margolis, University of California, Davis
Skyli McAfee, The Nature Conservancy (through 2017)
Beth Rose Middleton, University of California, Davis
Bradley H. Udall, Colorado State University

Technical Contributors
Mary E. Black, University of Arizona
Shallin Busch, National Oceanic and Atmospheric Administration
Brandon Goshi, Metropolitan Water District of Southern California

Review Editor
Cristina Bradatan, Texas Tech University

USGCRP Coordinators
Fredric Lipschultz, Senior Scientist and Regional Coordinator
Christopher W. Avery, Senior Manager

26. Alaska

Federal Coordinating Lead Author
Stephen T. Gray, U.S. Geological Survey

Chapter Lead
Carl Markon, U.S. Geological Survey (Retired)

Chapter Authors
Matthew Berman, University of Alaska, Anchorage
Laura Eerkes-Medrano, University of Victoria
Thomas Hennessy, U.S. Centers for Disease Control and Prevention
Henry P. Huntington, Huntington Consulting
Jeremy Littell, U.S. Geological Survey
Molly McCammon, Alaska Ocean Observing System
Richard Thoman, National Oceanic and Atmospheric Administration
Sarah Trainor, University of Alaska Fairbanks

Technical Contributors
Todd Brinkman, University of Alaska Fairbanks
Patricia Cochran, Alaska Native Science Commission
Jeff Hetrick, Alutiiq Pride Shellfish Hatchery
Nathan Kettle, University of Alaska Fairbanks
Robert Rabin, National Oceanic and Atmospheric Administration
Jacquelyn (Jaci) Overbeck, Alaska Department of Natural Resources
Bruce Richmond, U.S. Geological Survey
Ann Gibbs, U.S. Geological Survey
David K. Swanson, National Park Service
Todd Attwood, U.S. Geological Survey
Tony Fischbach, U.S. Geological Survey
Torre Jorgenson, Arctic Long Term Ecological Research
Neal Pastick, U.S. Geological Survey
Ryan Toohey, U.S. Geological Survey
Shad O'Neel, U.S. Geological Survey
Eran Hood, University of Alaska Southeast
Anthony Arendt, University of Washington
David Hill, Oregon State University
Lyman Thorsteinson, U.S. Geological Survey
Franz Mueter, University of Alaska Fairbanks
Jeremy Mathis, National Oceanic and Atmospheric Administration
Jessica N. Cross, National Oceanic and Atmospheric Administration
Jennifer Schmidt, University of Alaska Anchorage
David Driscoll, University of Virginia
Don Lemmen, Natural Resources Canada
Philip Loring, University of Saskatoon
Benjamin Preston, RAND Corporation
Stefan Tangen, University of Alaska Fairbanks
John Pearce, U.S. Geological Survey
Darcy Dugan, Alaska Ocean Observing System
Anne Hollowed, National Oceanic and Atmospheric Administration

Review Editor
Victoria Herrmann, The Arctic Institute

USGCRP Coordinators

Fredric Lipschultz, Senior Scientist and Regional Coordinator
Susan Aragon-Long, Senior Scientist

27. Hawai'i and U.S.-Affiliated Pacific Islands

Federal Coordinating Lead Author
David Helweg, DOI Pacific Islands Climate Adaptation Science Center

Chapter Lead
Victoria Keener, East-West Center

Chapter Authors
Susan Asam, ICF
Seema Balwani, National Oceanic and Atmospheric Administration
Maxine Burkett, University of Hawai'i at Mānoa
Charles Fletcher, University of Hawai'i at Mānoa
Thomas Giambelluca, University of Hawai'i at Mānoa
Zena Grecni, East-West Center
Malia Nobrega-Olivera, University of Hawai'i at Mānoa
Jeffrey Polovina, NOAA Pacific Islands Fisheries Science Center
Gordon Tribble, USGS Pacific Island Ecosystems Research Center

Technical Contributors
Malia Akutagawa, University of Hawai'i at Mānoa, Hawai'inuiākea School of Hawaiian Knowledge, Kamakakūokalani Center for Hawaiian Studies, William S. Richardson School of Law, Ka Huli Ao Center for Excellence in Native Hawaiian Law
Rosie Alegado, University of Hawai'i at Mānoa, Department of Oceanography, UH Sea Grant
Tiffany Anderson, University of Hawai'i at Mānoa, Geology and Geophysics
Patrick Barnard, U.S. Geological Survey–Santa Cruz
Rusty Brainard, NOAA Pacific Islands Fisheries Science Center
Laura Brewington, East-West Center, Pacific RISA
Jeff Burgett, Pacific Islands Climate Change Cooperative
Rashed Chowdhury, NOAA Pacific ENSO Applications Climate Center
Makena Coffman, University of Hawai'i at Mānoa, Urban and Regional Planning
Chris Conger, Sea Engineering, Inc.
Kitty Courtney, Tetra Tech, Inc.
Stanton Enomoto, Pacific Islands Climate Change Cooperative
Patricia Fifita, University of Hawai'ii, Pacific Islands Climate Change Cooperative
Lucas Fortini, USGS Pacific Island Ecosystems Research Center
Abby Frazier, USDA Forest Service
Kathleen Stearns Friday, USDA Forest Service, Institute of Pacific Islands Forestry

Neal Fujii, State of Hawai'i Commission on Water Resource Management
Ruth Gates, University of Hawai'i at Mānoa, School of Ocean and Earth Science and Technology
Christian Giardina, USDA Forest Service, Institute of Pacific Islands Forestry
Scott Glenn, State of Hawai'i Department of Health, Office of Environmental Quality Control
Matt Gonser, University of Hawai'i Sea Grant
Jamie Gove, NOAA Pacific Islands Fisheries Science Center
Robbie Greene, CNMI Bureau of Environmental and Coastal Quality
Shellie Habel, University of Hawai'i at Mānoa, School of Ocean and Earth Science and Technology
Justin Hospital, NOAA Pacific Islands Fisheries Science Center
Darcy Hu, National Park Service
Jim Jacobi, U.S. Geological Survey
Krista Jaspers, East-West Center, Pacific RISA
Todd Jones, NOAA Pacific Islands Fisheries Science Center
Charles Ka'ai'ai, Western Pacific Regional Fishery Management Council
Lauren Kapono, NOAA Papahānaumokuākea Marine National Monument
Hi'ilei Kawelo, Paepae O He'eia
Benton Keali'i Pang, U.S. Fish and Wildlife Service
Karl Kim, University of Hawai'i, National Disaster Preparedness Training Center
Jeremy Kimura, State of Hawai'i Commission on Water Resource Management
Romina King, University of Guam and Pacific Islands Climate Adaptation Science Center
Randy Kosaki, National Oceanic and Atmospheric Administration
Michael Kruk, ERT, Inc.
Mark Lander, University of Guam, Water and Environmental Research Institute
Leah Laramee, State of Hawai'i, Department of Land and Natural Resources
Noelani Lee, Ka Honua Momona
Sam Lemmo, State of Hawai'i Department of Land and Natural Resources, Interagency Climate Adaptation Committee
Rhonda Loh, Hawai'i Volcanoes National Park
Richard MacKenzie, USDA Forest Service, Institute of Pacific Islands Forestry
John Marra, National Oceanic and Atmospheric Administration
Xavier Matsutaro, Republic of Palau, Office of Climate Change
Marie McKenzie, Pacific Islands Climate Change Cooperative
Mark Merrifield, University of Hawai'i at Mānoa
Wendy Miles, Pacific Islands Climate Change Cooperative
Lenore Ohye, State of Hawai'i Commission on Water Resource Management
Kirsten Oleson, University of Hawai'i at Mānoa

Tom Oliver, University of Hawai'i at Mānoa, Joint Institute for Marine and Atmospheric Research
Tara Owens, University of Hawai'i Sea Grant
Jessica Podoski, U.S. Army Corps of Engineers—Fort Shafter
Dan Polhemus, U.S. Fish and Wildlife Service
Kalani Quiocho, NOAA Papahānaumokuākea Marine National Monument
Robert Richmond, University of Hawai'i, Kewalo Marine Lab
Joby Rohrer, O'ahu Army Natural Resources
Fatima Sauafea-Le'au, National Oceanic and Atmospheric Administration—American Sāmoa
Afsheen Siddiqi, State of Hawai'i, Department of Land and Natural Resources
Irene Sprecher, State of Hawai'i, Department of Land and Natural Resources
Joshua Stanbro, City and County of Honolulu Office of Climate Change, Sustainability and Resiliency
Mark Stege, The Nature Conservancy—Majuro
Curt Storlazzi, U.S. Geological Survey—Santa Cruz
William V. Sweet, National Oceanic and Atmospheric Administration
Kelley Tagarino, University of Hawai'i Sea Grant
Jean Tanimoto, National Oceanic and Atmospheric Administration
Bill Thomas, NOAA Office for Coastal Management
Phil Thompson, University of Hawai'i at Mānoa, Oceanography
Mililani Trask, Indigenous Consultants, LLC
Barry Usagawa, Honolulu Board of Water Supply
Kees van der Geest, United Nations University, Institute for Environment and Human Security
Adam Vorsino, U.S. Fish and Wildlife Service
Richard Wallsgrove, Blue Planet Foundation
Matt Widlansky, University of Hawai'i, Sea Level Center
Phoebe Woodworth-Jefcoats, NOAA Pacific Islands Fisheries Science Center
Stephanie Yelenik, USGS Pacific Island Ecosystems Research Center

Review Editor
Jo-Ann Leong, Hawai'i Institute of Marine Biology

USGCRP Coordinators
Allyza Lustig, Program Coordinator
Fredric Lipschultz, Senior Scientist and Regional Coordinator

28. Reducing Risks Through Adaptation Actions

Federal Coordinating Lead Authors
Jeffrey Arnold, U.S. Army Corps of Engineers
Roger Pulwarty, National Oceanic and Atmospheric Administration

Chapter Lead
Robert Lempert, RAND Corporation

Chapter Authors
Kate Gordon, Paulson Institute
Katherine Greig, Wharton Risk Management and Decision Processes Center at University of Pennsylvania (formerly New York City Mayor's Office of Recovery and Resiliency)
Cat Hawkins Hoffman, National Park Service
Dale Sands, Village of Deer Park, Illinois
Caitlin Werrell, The Center for Climate and Security

Technical Contributors
Lauren Kendrick, RAND Corporation
Pat Mulroy, Brookings Institution
Costa Samaras, Carnegie Mellon University
Bruce Stein, National Wildlife Federation
Tom Watson, The Center for Climate and Security
Jessica Wentz, Columbia University

Review Editor
Mary Ann Lazarus, Cameron MacAllister Group

USGCRP Coordinators
Sarah Zerbonne, Adaptation and Decision Science Coordinator
Fredric Lipschultz, Senior Scientist and Regional Coordinator

29. Reducing Risks Through Emissions Mitigation

Federal Coordinating Lead Author
Jeremy Martinich, U.S. Environmental Protection Agency

Chapter Lead
Jeremy Martinich, U.S. Environmental Protection Agency

Chapter Authors
Benjamin DeAngelo, National Oceanic and Atmospheric Administration
Delavane Diaz, Electric Power Research Institute
Brenda Ekwurzel, Union of Concerned Scientists
Guido Franco, California Energy Commission
Carla Frisch, U.S. Department of Energy
James McFarland, U.S. Environmental Protection Agency
Brian O'Neill, University of Denver (National Center for Atmospheric Research through June 2018)

Review Editor
Andrew Light, George Mason University

USGCRP Coordinators
David Reidmiller, Director
Christopher W. Avery, Senior Manager

Appendix 5. Frequently Asked Questions (FAQs)

Federal Coordinating Lead Author
David Reidmiller, U.S. Global Change Research Program

Lead Author
Matthew Dzaugis, U.S. Global Change Research Program/ICF

Contributing Authors
Christopher W. Avery, U.S. Global Change Research Program/ICF
Allison Crimmins, U.S. Environmental Protection Agency
LuAnn Dahlman, National Oceanic and Atmospheric Administration
David R. Easterling, NOAA National Centers for Environmental Information
Rachel Gaal, National Oceanic and Atmospheric Administration
Emily Greenhalgh, National Oceanic and Atmospheric Administration
David Herring, National Oceanic and Atmospheric Administration
Kenneth E. Kunkel, North Carolina State University
Rebecca Lindsey, National Oceanic and Atmospheric Administration
Thomas K. Maycock, North Carolina State University
Roberto Molar, National Oceanic and Atmospheric Administration
Brooke C. Stewart, North Carolina State University
Russell S. Vose, NOAA National Centers for Environmental Information

Technical Contributors
C. Taylor Armstrong, National Oceanic and Atmospheric Administration
Edward Blanchard-Wrigglesworth, University of Washington
James Bradbury, Georgetown Climate Center
Delavane Diaz, Electric Power Research Institute
Joshua Graff-Zivin, University of California, San Diego
Jessica Halofsky, University of Washington
Lesley Jantarasami, Oregon Department of Energy
Shannon LaDeau, Cary Institute of Ecosystem Studies
Elizabeth Marino, Oregon State University
Shaima Nasiri, U.S. Department of Energy
Matthew Neidell, Columbia University
Rachel Novak, U.S. Department of the Interior
Rick Ostfeld, Cary Institute of Ecosystem Studies
David Pierce, Scripps Institute of Oceanography
Catherine Pollack, National Oceanic and Atmospheric Administration
William V. Sweet, National Oceanic and Atmospheric Administration
Carina Wyborn, University of Montana
Laurie Yung, University of Montana–Missoula
Lewis Ziska, U.S. Department of Agriculture

USGCRP National Climate Assessment Staff

David Reidmiller, Director

Christopher W. Avery, Senior Manager

Bradley Akamine, Chief Digital Information Officer

Reuben Aniekwu, Global Change Information System Intern

Susan Aragon-Long, Senior Scientist

Natalie Bennett, Adaptation and Assessment Analyst

Ashley Bieniek-Tobasco, Health Program Coordinator

Mathia Biggs, Office Coordinator

Therese (Tess) S. Carter, Program Coordinator (until June 2017)

Apurva Dave, International Coordinator and Senior Analyst

David J. Dokken, Senior Program Officer

Matthew Dzaugis, Program Coordinator

Amrutha Elamparuthy, Data Manager

Anthony Flowe, Engagement and Communications Associate

Kristin Lewis, Senior Scientist

Fredric Lipschultz, Senior Scientist and Regional Coordinator

Allyza Lustig, Program Coordinator

Vincent O'Leary, Assessment Intern

Katie Reeves, Engagement and Communications Lead

Reid Sherman, Global Change Information System Lead

Mark Shimamoto, Program Coordinator (until August 2017)

Kathryn Tipton, Software Engineer

Katherine Weingartner, Program Assistant (until September 2017)

Sarah Zerbonne, Adaptation and Decision Science Coordinator

NOAA Technical Support Unit

David R. Easterling, NCA Technical Support Unit Director, NOAA National Centers for Environmental Information

Kenneth E. Kunkel, Lead Scientist, North Carolina State University

Sara W. Veasey, Creative Director, NOAA National Centers for Environmental Information

Brooke C. Stewart, Managing Editor and Lead Science Editor, North Carolina State University

Sarah M. Champion, Data Architect and Lead Information Quality Analyst, North Caronlina State University

Katharine M. Johnson, Web Developer and GIS Specialist, ERT, Inc.

James C. Biard, Software Engineer, North Carolina State University

Jessicca Griffin, Visual Communications Specialist and Graphic Designer, North Carolina State University

Angel Li, Web Developer, North Carolina State University

Thomas K. Maycock, Science Editor, North Carolina State University

Laura E. Stevens, Research Scientist, North Caronlina State University

Liqiang Sun, Research Scientist, North Carolina State University

Andrew Thrasher, Software Engineer, North Carolina State University

Andrea McCarrick, Editorial Assistant, North Carolina State University

Tiffany Means, Editorial Assistant, North Carolina State University

Andrew Buddenberg, Software Engineer, North Carolina State University (until October 2017)

Liz Love-Brotak, Graphic Designer, NOAA NCEI

Deborah Misch, Graphic Designer, TeleSolv Consulting

Deborah B. Riddle, Lead Graphic Designer (Report-in-Brief), NOAA NCEI

Mara Sprain, NCEI Librarian, LAC Group

Barbara Ambrose, Graphic Designer, Mississippi State University, Northern Gulf Institute

Andrew Ballinger, Research Scientist, North Carolina State University

Jennifer Fulford, Editorial Assistant, TeleSolv Consulting

Kristy Thomas, Metadata Specialist, ERT, Inc.

Terence R. Thompson, Climate Data Analyst, LMI

Caroline Wright, GIS Intern, North Carolina State University

Samantha Heitsch, Technical Writer, ICF

UNC Asheville's National Environmental Modeling and Analysis Center (NEMAC)

John Frimmel, Principal Software Developer

Ian Johnson, Geospatial and Science
Communications Associate

Karin Rogers, Director of Operations /
Research Scientist

This document responds to the requirements of Section 106 of the Global Change Research Act of 1990 (http://www.globalchange.gov/about/legal-mandate), and it meets all federal requirements associated with the *highly influential scientific assessment* (HISA) standard of the Information Quality Act (see Appendix 2: Information in the Fourth National Climate Assessment).

Appendix

Appendix

A5 Appendix 5. Frequently Asked Questions

This appendix is an update to the frequently asked questions (FAQs) presented in the Third National Climate Assessment (NCA3). New questions based on areas of emerging scientific inquiry are included alongside updated responses to the FAQs from NCA3. The answers are based on the U.S. Global Change Research Program's (USGCRP) sustained assessment products, other peer-reviewed literature, and consultation with experts.

Federal Coordinating Lead Author
David Reidmiller
U.S. Global Change Research Program

Lead Author
Matthew Dzaugis
U.S. Global Change Research Program/ICF

Contributing Authors
Christopher W. Avery
U.S. Global Change Research Program/ICF

Allison Crimmins
U.S. Environmental Protection Agency

LuAnn Dahlman
National Oceanic and Atmospheric Administration

David R. Easterling
NOAA National Centers for Environmental Information

Rachel Gaal
National Oceanic and Atmospheric Administration

Emily Greenhalgh
National Oceanic and Atmospheric Administration

David Herring
National Oceanic and Atmospheric Administration

Kenneth E. Kunkel
North Carolina State University

Rebecca Lindsey
National Oceanic and Atmospheric Administration

Thomas K. Maycock
North Carolina State University

Roberto Molar
National Oceanic and Atmospheric Administration

Brooke C. Stewart
North Carolina State University

Russell S. Vose
NOAA National Centers for Environmental Information

Technical Contributors are listed at the end of the chapter.

Recommended Citation for Chapter
Dzaugis, M.P., D.R. Reidmiller, C.W. Avery, A. Crimmins, L. Dahlman, D.R. Easterling, R. Gaal, E. Greenhalgh, D. Herring, K.E. Kunkel, R. Lindsey, T.K. Maycock, R. Molar, B.C. Stewart, and R.S. Vose, 2018: Frequently Asked Questions. In *Impacts, Risks, and Adaptation in the United States: Fourth National Climate Assessment, Volume II* [Reidmiller, D.R., C.W. Avery, D.R. Easterling, K.E. Kunkel, K.L.M. Lewis, T.K. Maycock, and B.C. Stewart (eds.)]. U.S. Global Change Research Program, Washington, DC, USA, pp. 1435–1506. doi: 10.7930/NCA4.2018.AP5

On the Web: https://nca2018.globalchange.gov/chapter/appendix-5

Contents

Introduction to climate change 192

How do we know Earth is warming?.. 192

What makes recent climate change different from warming in the past?..................... 193

What's the difference between global warming and climate change?.......................... 195

Climate Science 196

What are greenhouse gases, and what is the greenhouse effect?.............................. 196

Why are scientists confident that human activities are the primary cause of recent climate change?... 198

What role does water vapor play in climate change?.. 201

How are El Niño and climate variability related to climate change?............................ 202

Temperature and Climate Projections....................................... 204

What methods are used to record global surface temperatures and measure changes in climate?....... 204

Were there predictions of global cooling in the 1970s?.. 206

How are temperature and precipitation patterns projected to change in the future?................. 207

How do computers model Earth's climate?.. 209

Can scientists project the effects of climate change for local regions?........................ 211

What are key uncertainties when projecting climate change? 212

Is it getting warmer everywhere at the same rate?... 215

What do scientists mean by the "warmest year on record"? 218

How do climate projections differ from weather predictions? 219

Climate, Weather, and Extreme Events 221

Was there a "hiatus" in global warming?.. 221

What is an extreme event?... 223

Have there been changes in extreme weather events? 224

Can specific weather or climate-related events be attributed to climate change?.................. 226

Could climate change make Atlantic hurricanes worse?.. 227

Societal Effects 229

How is climate change affecting society?..229

What is the social cost of carbon?..231

What are climate change mitigation, adaptation, and resilience?...232

Is timing important for climate mitigation?..233

Are there benefits to climate change?...235

Are some people more vulnerable to climate change than others?..236

How will climate change impact economic productivity?...237

Can we slow climate change?...238

Can geoengineering be used to remove carbon dioxide from the atmosphere
or otherwise reverse global warming?...239

Ecological Effects 240

What causes global sea level rise, and how will it affect coastal areas in the coming century?...............240

How does global warming affect arctic sea ice cover? 242

Is Antarctica losing ice? What about Greenland?...245

How does climate change affect mountain glaciers?..246

How are the oceans affected by climate change?...247

What is ocean acidification, and how does it affect marine life? 249

How do higher carbon dioxide concentrations affect plant communities and crops?..............251

Is climate change affecting U.S. wildfires?...252

Does climate change increase the spread of mosquitoes or ticks?...254

References 256

Introduction to Climate Change

How do we know Earth is warming?

Many indicators show conclusively that Earth has warmed since the 19th century. In addition to warming shown in the observational record of oceanic and atmospheric temperature, other evidence includes melting glaciers and continental ice sheets, rising global sea level, a longer frost-free season, changes in temperature extremes, and increases in atmospheric humidity, all consistent with long-term warming.

Observations of surface temperature taken over Earth's land and ocean surfaces since the 19th century show a clear warming trend. Temperature observations have been taken consistently since the 1880s or earlier at thousands of observing sites around the world. Additionally, instruments on ships, buoys, and floats together provide a more-than-100-year record of sea surface temperature showing that the top 6,500 feet of Earth's ocean is warming in all basins.[1] These observations are consistent with readings from satellite instruments that measure atmospheric and sea surface temperatures from space. Used together, land-, ocean-, and space-based temperature observations show clear evidence of warming at Earth's surface over climatological timescales (http://www.globalchange.gov/browse/indicators for more indicators of change) (see also Ch. 2: Climate).

Scientists around the world have been measuring the extent and volume of ice contained in the same glaciers every few years since 1980. These measurements show that, globally, there is a large net volume loss in glacial ice since the 1980s. However, the rate of the ice loss varies by region, and in some cases yearly glacier advances are observed (see FAQ "How does climate change affect mountain glaciers?"). Ice sheets on Antarctica and Greenland have been losing ice mass consistently since 2002, when advanced satellite measurements of their continental ice mass began (see FAQ "Is Antarctica losing ice? What about Greenland?"). Arctic sea ice coverage has been monitored using satellite imagery since the late 1970s, showing consistent and large declines in September, the time of year when the minimum coverage occurs.[2]

There are additional observational lines of evidence for warming. For example, the area of land in the Northern Hemisphere covered by snow each spring is now smaller on average than it was in the 1960s.[3] Tide gauges and satellites show that global sea level is rising, both as a result of the addition of water to the ocean from melting glaciers and from the expansion of seawater as it warms (Ch. 2: Climate; Ch. 8: Coastal). Lastly, as air warms, its capacity to hold water vapor increases, and measurements show that atmospheric humidity is increasing around the globe, consistent with a warming climate (see Ch. 3: Water; see also Ch. 1: Overview, Figure 1.2 for more indicators of a warming world).

What makes recent climate change different from warming in the past?

Increases in global temperature since the 1950s are unusual for two reasons. First, current changes are primarily the result of human activities rather than natural physical processes. Second, temperature changes are occurring much faster than they did in the past.

Our planet's climate has changed before. Sedimentary rocks and fossils show clear evidence for a series of long cold periods—called ice ages—followed by warm periods. Common archaeological and geological processes for dating past events show that these cycles of cooling and warming occurred about once every 100,000 years for at least the last million years.

Before major land-use changes and industrialization, changes in global temperature were caused by natural factors, including regular changes in Earth's orbit around the sun, volcanic eruptions, and changes in energy from the sun.[4] Major warming and cooling events were driven by natural variations of Earth's orbit that altered the amount of sunlight reaching Earth's Arctic and Antarctic regions, resulting in the retreat and advance of massive ice sheets. Additionally, quiescent or active periods of volcanic eruptions also could contribute to warming or cooling events, respectively.[5]

Natural factors are still affecting the planet's climate today (see Figure A5.5). Yet since the beginning of the Industrial Revolution, human use of coal, oil, and gas has rapidly changed the composition of the atmosphere (Figure A5.1). Land-use changes (such as deforestation), cement production, and animal production for food have also contributed to the increase in levels of greenhouse gases in the atmosphere. Unlike past changes in climate, today's warming is driven primarily by human activity rather than by natural physical processes (see Figure A5.5) (see also Ch. 2: Climate).

Current warming is also happening much faster than it did in the past. Scientific records from ice cores, tree rings, soil boreholes, and other "natural thermometers"—often called proxy climate data—show that the recent increase in temperature is unusually rapid compared to past changes (see Figures A5.2 and A5.4). After an ice age, Earth typically took thousands of years to warm up again; the observed rate of warming over the last 50 years is about eight times faster than the average rate of warming from a glacial maximum to a warm interglacial period.[4]

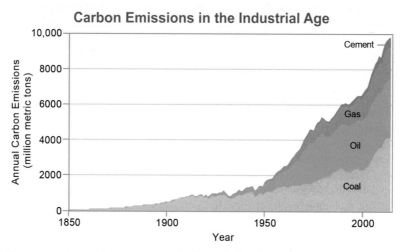

Figure A5.1 Humans have changed the atmosphere by burning coal, oil, and gas for energy and by producing cement. This graph shows the total global carbon emissions from these activities from 1850 to 2009. A range of other human activities, such as cutting down forests and livestock production, account for additional carbon emissions. Source: Walsh et al. 2014.[6]

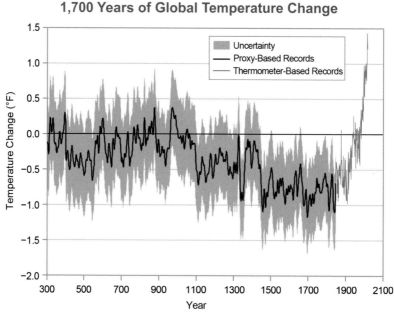

Figure A5.2 Average global temperature has increased rapidly over the last 1,700 years compared to the 1961–1990 average. The red line shows temperature data based on surface observations. The black line shows temperature data from proxies, including data from tree rings, ice cores, corals, and marine sediments. The comparison of proxy- and thermometer-based records suggests that temperatures are now higher than they have been in at least 1,700 years. The steep portion of the graph since about 1950 shows how rapidly temperature has increased compared to previous changes. Source: adapted from Mann et al. 2008.[7]

What's the difference between global warming and climate change?

Though some people use the terms "global warming" and "climate change" interchangeably, their meanings are slightly different. Global warming refers only to Earth's rising surface temperature, while climate change includes temperature changes and a multitude of effects that result from warming, including melting glaciers, increased humidity, heavier rainstorms, and changes in the patterns of some climate-related extreme events.

By itself, the phrase global warming refers to increases in Earth's annual average surface temperature. Today, however, when people use the phrase, they usually mean the recent warming that is due in large part to the rapid increase of greenhouse gases (GHGs) in the atmosphere from human activities such as deforestation and the burning of fossil fuels for energy. Thus, "global warming" has become a form of shorthand for a complex scientific process.

The entire globe is not warming uniformly. Some areas may cool (such as the North Atlantic Ocean), while some may warm faster than the global average (such as the Arctic). The term climate change refers to the full range of consequences or impacts that occur as atmospheric levels of GHGs rise and different parts of the earth system respond to a higher average surface temperature. For instance, observed long-term trends, such as increases in the frequency of drought and heavy precipitation events, are not technically warming trends, but they are related to current warming and are processes of climate change (Ch 2: Climate).

Climate Science

What are greenhouse gases, and what is the greenhouse effect?

Greenhouse gases (GHGs) are gases that absorb and emit thermal (heat) infrared radiation. Carbon dioxide, methane, nitrous oxide, ozone, and water vapor are the most prevalent GHGs in Earth's atmosphere. These gases absorb heat emitted by Earth's surface and re-emit that heat into Earth's atmosphere, making it much warmer than it would be otherwise—a process known as the greenhouse effect.

Most of Earth's atmosphere is made up of nitrogen (N_2) and oxygen (O_2), neither of which is considered a greenhouse gas. Other gases, known as greenhouse gases (GHGs), behave very differently from O_2 and N_2 when it comes to infrared radiation emitted from Earth. GHGs, such as water vapor, carbon dioxide (CO_2), and methane (CH_4), have a more complex molecular structure (made up of three or more atoms, as opposed to the symmetrical, two-atom molecules of O_2 and N_2) that absorbs some of the energy emitted from Earth's surface and then re-radiates that energy in all directions, including back down towards the surface. This ultimately traps energy in the lower atmosphere in the form of heat (Figure A5.3). This greenhouse effect makes the average temperature of Earth nearly 60°F warmer than it would be in the absence of these GHGs. Even a tiny amount of these gases can have a huge effect on the amount of heat trapped in the lower atmosphere, just like a tiny amount of anthrax can have a huge effect on human health.

Many GHGs, including CO_2, CH_4, water vapor, and nitrous oxide (N_2O), occur naturally in the atmosphere. However, atmospheric concentrations of these GHGs have been rising over the last few centuries as a result of human activities. In addition, human activities have added new, entirely human-made GHGs to the atmosphere, including chlorofluorocarbons (CFCs), hydrofluorocarbons (HFCs), perfluorocarbons (PFCs), and sulfur hexafluoride (SF6).[5]

As the global population has increased, so have GHG emissions. This in turn makes the greenhouse effect stronger, resulting in higher average temperature around the globe (Ch 2: Climate).

The Greenhouse Effect

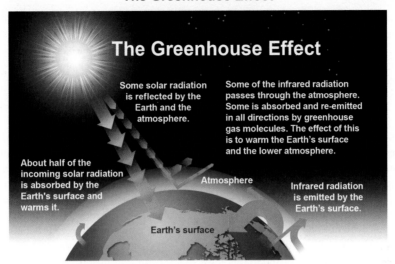

Figure A5.3: The figure shows a simplified representation of the greenhouse effect. About half of the sun's radiation reaches Earth's surface, while the rest is reflected back to space or absorbed by the atmosphere. Naturally occurring greenhouse gases, including carbon dioxide (CO_2), methane (CH_4), and nitrous oxide (N_2O), do not absorb most of the incoming shortwave (visible) energy from the sun, but they do absorb the longwave (infrared) energy re-radiated from Earth's surface. This energy is then re-emitted in all directions, keeping the surface of the planet much warmer than it would be otherwise. Human activities—predominantly the burning of fossil fuels (coal, oil, and gas)—are increasing levels of CO_2 and other GHGs in the atmosphere, which is amplifying the natural greenhouse effect and thus increasing Earth's temperature. Source: adapted from EPA 2016.[8]

> ### Why are scientists confident that human activities are the primary cause of recent climate change?
>
> Many independent lines of evidence support the finding that human activities are the dominant cause of recent (since 1950) climate change. These lines of evidence include changes seen in the observational records that are consistent with our understanding, based on physics, of how the climate system should change due to human influences. Other evidence comes from climate modeling studies that closely reproduce the observed temperature record.

The Climate Science Special Report[9] concludes, "human activities, especially emissions of greenhouse gases, are the dominant cause of the observed warming since the mid-20th century." The Earth's climate only warms or cools significantly in response to changes that affect the balance of incoming and outgoing energy. Over long timescales (tens to hundreds of thousands of years), orbital cycles produce long periods of warming and cooling. Over shorter timescales, two factors could generally force changes in Earth's temperature to a measurable degree: (1) changes in the amount of energy put out by the sun, and (2) changes in the concentrations of greenhouse gases (GHGs) in Earth's atmosphere. Recent measurements of the sun's energy show no trend over the last 50 years. Additionally, observations show that the lower atmosphere (troposphere) has warmed while the upper atmosphere (stratosphere) has cooled. If the observed warming had been due to an increase in energy from the sun, then all layers of Earth's atmosphere would have warmed, which is not what scientists observe. Thus, we can eliminate changes in the energy received from the sun as a major factor in the warming observed since about 1950.[10]

This leaves the possibility that changes in GHG concentrations in the atmosphere are the primary cause of recent warming. Atmospheric carbon dioxide (CO_2) levels have increased from approximately 270 parts per million (ppm) during preindustrial times to the current 408 ppm observed in 2018 (see https://www.esrl.noaa.gov/gmd/ccgg/trends/)—levels that exceed any observed over the past 800,000 years (Figure A5.4). In addition, atmospheric concentrations of other GHGs (including methane and nitrous oxide) have increased over the same period. This increase in GHG concentrations has coincided with the observed increase in global temperature. Scientists use methods that provide chemical "fingerprints" of the source of these increased emissions and have shown that the 40% increase in atmospheric CO_2 levels since the Industrial Revolution is due mainly to human activities (primarily the combustion of fossil fuels) and not due to natural carbon cycle processes.[5]

Other evidence attributing human activities as the dominant driver of observed warming comes from climate modeling studies. Computer simulations of Earth's climate based on historical data of observed changes in natural and human influences accurately reproduce the observed temperature record over the last 120 years. These results show that without human influences, such as the observed increases in GHG emissions, Earth's surface would have cooled slightly over the past half century. The only way to closely replicate the observed warming is to include both natural and human forcing changes in climate models (Figure A5.5). Thus, the observational record and modeling studies both point to human factors being the main cause for the recent warming (Ch.2: Climate).

800,000 Years of CO₂ and Temperature Change

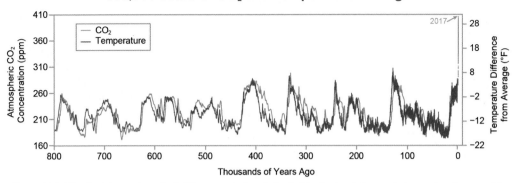

Figure A5.4: This chart shows atmospheric CO_2 concentrations (left axis, blue line) and changes in temperature (compared to the average over the last 1,000 years; right axis, red line) over the past 800,000 years, as recorded in ice cores from Antarctica. Also shown are modern instrumental measurements of CO_2 concentrations through 2017. Current CO_2 concentrations are much higher than any levels observed over the past 800,000 years. Source: adapted from EPA 2017.[11]

Human and Natural Influences on Global Temperature

Figure A5.5: Both human and natural factors influence Earth's climate, but the long-term global warming trend observed over the past century can only be explained by the effect that human activities have had on the climate.

Sophisticated computer models of Earth's climate system allow scientists to explore the effects of both natural and human factors. In all three panels of this figure, the black line shows the observed annual average global surface temperature for 1880–2017 as a difference from the average value for 1880–1910.

The top panel (a) shows the temperature changes simulated by a climate model when only natural factors (yellow line) are considered. The other lines show the individual contributions to the overall effect from observed changes in Earth's orbit (brown line), the amount of incoming energy from the sun (purple line), and changes in emissions from volcanic eruptions (green line). Note that no long-term trend in globally averaged surface temperature over this time period would be expected from natural factors alone.[4]

The middle panel (b) shows the simulated changes in global temperature when considering only human influences (dark red line), including the contributions from emissions of greenhouse gases (purple line) and small particles (referred to as aerosols, brown line) as well as changes in ozone levels (orange line) and changes in land cover, including deforestation (green line). Changes in aerosols and land cover have had a net cooling effect in recent decades, while changes in near-surface ozone levels have had a small warming effect.[5] These smaller effects are dominated by the large warming influence of greenhouse gases such as carbon dioxide and methane. Note that the net effect of human factors (dark red line) explains most of the long-term warming trend.

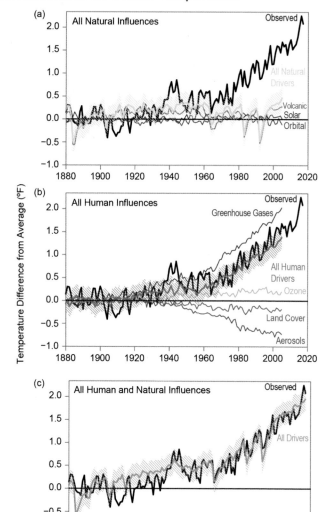

The bottom panel (c) shows the temperature change (orange line) simulated by a climate model when both human and natural influences are included. The result matches the observed temperature record closely, particularly since 1950, making the dominant role of human drivers plainly visible.

Researchers do not expect climate models to exactly reproduce the specific timing of actual weather events or short-term climate variations, but they do expect the models to capture how the whole climate system behaves over long periods of time. The simulated temperature lines represent the average values from a large number of simulation runs. The orange hatching represents uncertainty bands based on those simulations. For any given year, 95% of the simulations will lie inside the orange bands. See Chapter 2: Climate for more information. Source: NASA GISS.

What role does water vapor play in climate change?

Water vapor is the most abundant greenhouse gas (GHG) in the atmosphere and plays an important role in Earth's climate, significantly increasing Earth's temperature. However, unlike other GHGs, water vapor can condense and precipitate, so water vapor has a short life span in the atmosphere. Air temperature, and not emissions, controls the amount of water vapor in the lower atmosphere. For this reason, water vapor is considered a feedback agent and not a driver of climate change.

Water vapor is the primary GHG in the atmosphere, and its contribution to Earth's greenhouse effect is about two or three times that of carbon dioxide (CO_2). Human activities directly add water vapor to the atmosphere primarily through increasing evaporation from irrigation, power plant cooling, and combustion of fossil fuels. Other GHGs, such as CO_2, are not condensable at atmospheric temperatures and pressures, so they will continue to build up in the atmosphere as long as their emissions continue.[12]

The amount of water vapor in the lower atmosphere (troposphere) is mainly controlled by the air temperature and proximity to a water source, such as an ocean or large lake, rather than by emissions from human activities. Fluctuations in air temperature change the amount of water vapor that the air can hold, with warmer air capable of holding more moisture. Increases in water vapor levels in the lower atmosphere are considered a "positive feedback" (or self-reinforcing cycle) in the climate system. As increasing concentrations of other GHGs (for example, carbon dioxide, methane, and nitrous oxide) warm the atmosphere, atmospheric water vapor concentrations increase, thereby amplifying the warming effect (Figure A5.6). If atmospheric concentrations of CO_2 and other GHGs decreased, air temperature would drop, decreasing the ability of the atmosphere to hold water vapor, further decreasing temperature.[5,12]

Water Vapor and the Greenhouse Effect

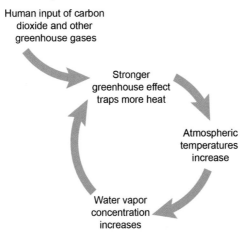

Figure A5.6: As emissions of carbon dioxide and other greenhouse gases increase, the strength of the greenhouse effect increases, which drives an increase in global temperature. This in turn increases the amount of water vapor in the lower atmosphere. Because water vapor is itself a greenhouse gas, the increase in atmospheric water vapor can further strengthen the greenhouse effect. Source: USGCRP.

How are El Niño and climate variability related to climate change?

El Niño and other forms of natural climate variability are not caused by humans, but their frequency, duration, extent, or intensity might be affected by greenhouse gas emissions from human activities. Natural climate variability produces short-term regional changes in temperature and weather patterns, whereas human-caused climate change is a persistent, long-term phenomenon.

Climate variability refers to the natural changes in climate that fall within the observed range of extremes for a particular region, as measured by temperature, precipitation, and frequency of events. Drivers of climate variability include the El Niño–Southern Oscillation (ENSO) and other phenomena. ENSO is a quasi-periodic warming or cooling of the of the sea surface temperatures in the tropical eastern Pacific and is often referred to by its phase of El Niño (warm phase) or La Niña (cool phase). These different ENSO phases can have varying ecosystem and economic effects, especially in certain fishing communities, while also influencing weather worldwide (Figure A5.7). In the United States, El Niño conditions generally correspond with warmer than average sea surface and air temperatures along the West Coast, wetter conditions in the Southwest, cooler temperatures in the Southeast, and warmer conditions in the Northeast. In contrast, the La Niña phase of ENSO corresponds to cooler temperature in the U.S. Northwest and dryer and warmer conditions in the Southeast, along with increased upwelling along the West Coast.

Evidence from paleoclimate records suggests that there have been changes in the frequency and intensity of ENSO events in the past. Human-caused climate change might also affect the frequency and magnitude of ENSO events and can exacerbate or ameliorate regional ENSO impacts. For example, if there is a strong La Niña event that results in dry conditions in the Southwest, those conditions may be exacerbated by additional drying due to climate change. ENSO is a complex phenomenon, but new research is shedding light on the many factors influencing how climate change affects the ENSO cycle.[13]

El Niño/La Niña Cause Short-Term Changes in Weather Patterns

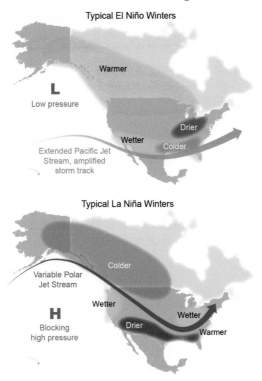

Figure A5.7: El Niño and La Niña events create different weather patterns during winters (January through March) over North America. (top) During an El Niño, there is a tendency for a strong jet stream and storm track across the southern part of the United States. The southern tier of Alaska and the U.S. Pacific Northwest tend to be warmer than average, whereas the southern United States tends to be cooler and wetter than average. (bottom) During a La Niña, there is a tendency for very wave-like jet stream flow over the United States and Canada, with colder and stormier than average conditions across the North and warmer and less stormy conditions across the South. Source: Perlwitz et al. 2017.[13]

Temperature and Climate Projections

What methods are used to record global surface temperatures and measure changes in climate?

Global surface temperatures are measured by using data from weather stations over land and by ships and buoys over the ocean. Global surface temperature records date back more than 300 years in some locations, and near-global coverage has existed since the late 1800s. Multiple research groups have examined U.S. and global temperature records in great detail, taking into account changes in instruments, the time of observations, station location, and any other potential sources of error. Although there are slight differences among datasets—due to choices in data selection, analysis, and averaging techniques—these differences do not change the clear result that global surface temperature is rising.

Climate change is best measured by assessing trends over long periods of time (generally greater than 30 years), which means we need global surface temperature records that include data from before the satellite age. Scientists who obtain, digitize, and collate long-term temperature records take great care to ensure that any potentially skewed measurements—such as a change in instrument method or location or a change in the time of day a recording is made—do not affect the integrity of the dataset. Researchers rigorously examine the data to identify and adjust for any such effects before using it to evaluate long-term climate trends. Different choices in data selection, analysis, and averaging techniques by multiple independent research teams mean that each dataset varies slightly. Even with these variations, however, multiple independently produced results are in very good agreement at both global and regional scales: all global surface temperature datasets indicate that the vast majority of Earth's surface has warmed since 1901 (Figure A5.8).

Scientists also consider other influences that could impact temperature records, such as whether data from thermometers located in cities are skewed by the urban heat island effect, where heat absorbed by buildings and asphalt makes cities warmer than the surrounding countryside. When determining climate trends, data corrections to these temperature records have adequately accounted for this effect. At the global scale, evidence of global warming over the past 50 years is still observed even if all of the urban stations are removed from the global temperature record. Studies have also shown that the warming trends of rural and urban areas that are in close proximity essentially match, even though the urban areas may have higher temperatures overall.[14]

Global Temperature Increase Shown in Multiple Datasets

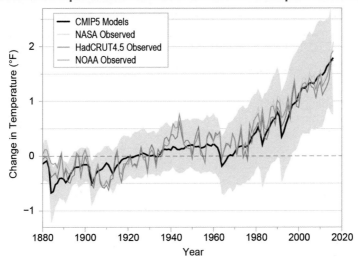

Figure A5.8: This chart shows observations of global annual average temperatures from three different datasets—one from NASA (yellow line), one from NOAA (orange line), and one from the University of East Anglia in conjunction with the United Kingdom's Met Office (HadCRUT4.5, brown line)—along with historical simulations of global temperature from the Coupled Model Intercomparison Project Phase 5 (CMIP5) ensemble of climate models (black line). The lines show annual differences in temperature relative to the 1901–1960 average. Small differences among datasets, due to choices in data selection, analysis, and averaging techniques, do not affect the conclusion that global surface temperatures are increasing. Source: adapted from Knutson et al. 2016.[15]

> ### Were there predictions of global cooling in the 1970s?
>
> No. A review of the scientific literature from the 1970s shows that the broad climate science community did not predict "global cooling" or an "imminent" ice age. On the contrary, even then, discussions of human-related warming dominated scientific publications on climate and human influences.

Scientific understanding of what are called the Milankovitch cycles (cyclical changes in Earth's orbit that can explain the onset and ending of ice ages) led a few scientists in the 1970s to contemplate that the current warm interglacial period might be ending soon, leading to a new ice age over the next few centuries. These few speculations were picked up and amplified by the media. But at that time there were far more scientific articles describing how warming would occur from the increase in atmospheric concentrations of greenhouse gases from human activities, including the burning of fossil fuels (Figure A5.9). The latest information suggests that if Earth's climate was being controlled primarily by natural factors, the next cooling cycle would begin sometime in the next 1,500 years. However, humans have so altered the composition of the atmosphere that the next ice age has likely now been delayed. That delay could potentially be tens of thousands of years.[6]

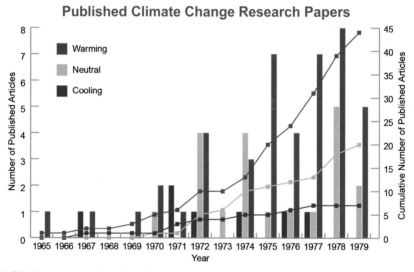

Published Climate Change Research Papers

Figure A5.9: This chart compares the number of papers classified as predicting, implying, or providing supporting evidence for future global cooling, warming, and neutral categories published from 1965 to 1979. The bars indicate the number of articles published per year. The lines with squares indicate the cumulative number of articles published. Over this period the literature survey found 7 papers suggesting future cooling (blue line), 20 neutral (yellow line), and 44 warming (red line). Source: Peterson et al. 2008.[16]

How are temperature and precipitation patterns projected to change in the future?

Our world will continue to warm in the future because of historic emissions of greenhouse gases (GHGs), but the amount of warming will depend largely on the level of future emissions of GHGs and the choices humans make. If humans continue burning fossil fuels at or above our current rate through the end of the century, scientists project Earth will warm about 9°F, relative to preindustrial times (prior to 1750). Precipitation is projected to still be seasonally and regionally variable, but on average, projections show high-latitude areas getting wetter and subtropical areas getting drier. The frequency and intensity of very heavy precipitation are expected to increase, increasing the likelihood of flooding. Climate change will not affect all places in the same way or to the same degree but will vary at regional levels.

In the coming decades, scientists project that global average temperature will continue to increase (Ch. 2: Climate), although natural variability will continue to play a significant role in year-to-year changes. Sizeable variations from global average changes are possible at the regional level. Even if humans drastically reduce levels of GHG emissions, near-term warming will still occur because there is a lag in the temperature response to changes in atmospheric composition (Figure A5.10).

Over the next couple decades, natural variability and the response of Earth's climate system to historic emissions will be the primary determinants of observed warming. After about 2050, however, the rate and amount of emissions of GHGs released by human activities, as well as the response of Earth's climate system to those emissions, will be the primary determining factors in changes in global and regional temperature (Figure A5.13) (see also Ch. 2: Climate). Efforts to rapidly and significantly reduce emissions of GHGs can still limit the global temperature increase to 3.6°F (2°C) by the end of the century relative to preindustrial levels.[17]

Precipitation patterns are also expected to continue to change throughout this century and beyond. The trends observed in recent decades are expected to continue, with more precipitation projected to fall in the form of heavier precipitation events.[3] Such events increase the likelihood of flooding, even in drought-prone areas. As with increases in global average temperature, large-scale shifts towards wetter or drier conditions and the projected increases in heavy precipitation are expected to be greater under higher GHG emissions scenarios (for example, RCP8.5) versus lower ones (for example, RCP4.5). Projected warming is also expected to lead to an increase in the fraction of total precipitation falling as rain rather than snow, which reduces snowpack on the margins of areas that now have reliable snowpack accumulation during the cold season (see, for example, Ch. 24: Northwest, KM 2).

Observed and Projected Changes in Global Temperature

Global Average Temperature Change

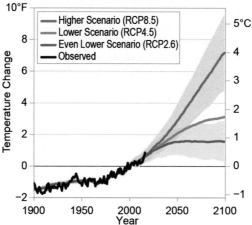

Figure A5.10: This figure shows both observed and projected changes in global average temperature. Under a representative concentration pathway (RCP) consistent with a higher scenario (RCP8.5; red) by 2080–2099, global average temperature is projected to increase by 4.2°–8.5°F (2.4°–4.7°C; burnt orange shaded area) relative to the 1986–2015 average. Under a lower scenario (RCP4.5; blue) global average temperature is projected to increase by 1.7°–4.4°F (0.9°–2.4°C; range not shown on graph) relative to 1986–2015. Under an even lower scenario (RCP2.6; green) temperature increases could be limited to 0.4°–2.7°F (0.2°–1.5°C; green shaded area) relative to 1986–2015. Limiting the rise in global average temperature to less than 2.2°F (1.2°C) relative to 1986–2015 is approximately equivalent to 3.6°F (2°C) or less relative to preindustrial temperatures. Thick lines within shaded areas represent the average of multiple climate models. The shaded regions illustrate the 5% to 95% confidence intervals for the respective projections. Source: adapted from Wuebbles et al. 2017.[4]

How do computers model Earth's climate?

Global climate models enable scientists to create "virtual Earths," where they can analyze caus-es and effects of past changes in temperature, precipitation, and other climate variables. Today's climate models can accurately reproduce broad features of past and present climate, such as the location and strength of the jet stream, the spatial distribution and seasonal cycle of precipitation, and the natural occurrence of extreme weather events, such as heat and cold waves, droughts and floods, and hurricanes. They also can reproduce historic natural cycles, such as the periodic occur-rence of ice ages and interglacial warm periods, as well as the human-caused warming that has occurred over the last 50 years. While uncertainties remain, scientists have confidence in model projections of how climate is likely to change in the future in response to key variables, such as an increase in human-caused emissions of greenhouse gases, in part because of how accurately they can represent past climate changes.

Climate models are based on equations that represent fundamental laws of nature and the many processes that affect Earth's climate system. By dividing the atmosphere, land, and ocean into smaller spatial units to solve the equations, climate models capture the evolving patterns of atmo-spheric pressures, winds, temperatures, and precipitation. Over longer time frames, these models simulate wind patterns, high- and low-pressure systems, ocean currents, ice and snowpack accumulation and melting, soil moisture, extreme weather occurrences, and other environmental characteristics that make up the climate system (Figure A5.11).[18]

Some important processes, including cloud formation and atmospheric mixing, are represented by approximate relationships, either because the processes are not fully understood or they are at a scale that a model cannot directly represent. These approximations lead to uncertainties in model simulations of climate. Approximations are not the only uncertainties associated with climate models, as discussed in the FAQ "What are key uncertainties when projecting climate change?"

Comparison of Climate Models and Observed Temperature Change

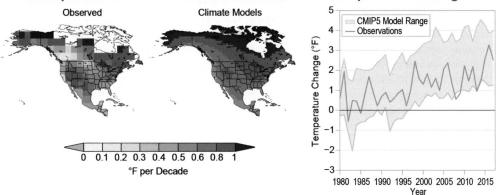

Figure A5.11: Climate simulations (right map) can capture the approximate geographical patterns and magnitude of the surface air temperature trend seen in observational data for the period 1980–2017 (left map). The warming pattern seen in the right map is an average based on 43 different global climate models from the Coupled Model Intercomparison Project Phase 5 (CMIP5). The graphical representation shows the range of temperature changes simulated by the models for North America (relative to 1901–1960; gray shading, 5th to 95th percentile range) overlaid by the observed annual average temperatures over North America (orange line). The observed temperature changes are a result of both human contributions to recent warming and natural temperature variations. Averaging the simulations from multiple models suppresses the natural variations and thus shows mainly the human contribution, which is part of the reason small-scale details are different between the two maps. Sources: (maps) adapted from Walsh et al. 2014[6] (and graph) NOAA NCEI and CICS-NC.

Can scientists project the effects of climate change for local regions?

Yes, though there are limitations. With advances in computing power, the future effects of climate change can be projected more accurately for local communities. Local high-resolution (down-scaled) climate modeling can be used to produce data at a scale of 1–20 miles. These downscaled projections show climate-related impacts at the local level and can be an important tool for community planners and decision-makers.

One significant research focus recently has been to develop models of climate impacts on a relatively small geographic scale. Most global climate projections use grid units that may be too coarse to properly represent mountains, coastlines, and other important features of a local landscape. Recently, two different approaches have been used by scientists to project local climate conditions.

The first is a statistical approach that uses local observations in conjunction with global models to project future changes. The local observations required for this approach are available only for limited regions and for a few climate variables (mainly temperature and precipitation; Figure A5.12).

The second method is a so-called dynamical approach that uses an additional high-resolution computer model—similar to a weather prediction model— to account for complex topography and varying land cover that can impact climate on the local level. High-resolution dynamical models are complete enough to simulate numerous climate variables (temperature, precipitation, winds, humidity, surface sunlight, etc.) and do not require the local observations required for the statistical approach. However, these models require an immense amount of computing power. Today's most powerful supercomputers enable climate scientists to examine the effects of climate change in ways that were impossible just five years ago. Over the next decade, computer speeds are predicted to increase 100-fold or more, improving climate projections and models on both the global and local levels.

It should also be noted that both statistical and dynamical approaches have biases and errors that, when combined with uncertainties from global model simulations, can reduce the level of confidence in these more localized projections (see Hayhoe et al. 2017[18] for more details).

Climate Modeling for Smaller Regions

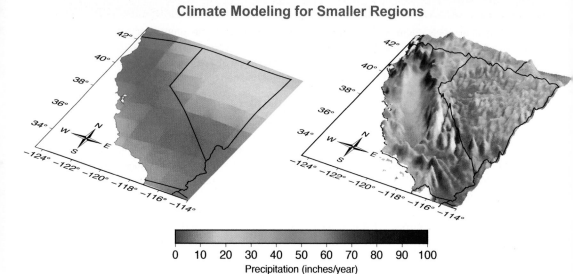

Figure A5.12: The figure shows projections of annual precipitation (in inches) in California and Nevada in a global climate model with a resolution of 100 miles (left) and, after using a statistical model to account for the effects of topography, at a resolution of 3.6-miles (right). The global model has only a few grid cells over the entire state of California, so it does not resolve the coastal mountain range, interior valley, or Sierra Nevada on the border with Nevada. The precipitation field in the right panel, by contrast, captures the wet conditions on the west slopes of the mountains and the dry, rain shadow region to the east of the mountains. The topography has been exaggerated for clarity and by the same amount in both panels. Source: UCSD Scripps Institute of Oceanography.

What are key uncertainties when projecting climate change?

The precise amount of future climate change that will occur over the rest of this century is uncertain, mainly due to uncertainties in emissions, natural variability, and differences in scientific models.

First, projections of future climate changes are usually based on scenarios (or sets of assumptions) regarding how future emissions may change due to changes in population, energy use, technology, and economics. Society may choose to reduce emissions or continue on a pathway of increasing emissions. The differences in projected future climate under different scenarios are generally small for the next few decades. By the second half of the century, however, human choices, as reflected in these scenarios, become the key determinant of future climate change (Figure A5.13).

A second source of uncertainty is natural variability, which affects the climate over timescales from months to decades. These natural variations are largely unpredictable, such as a volcanic eruption, and are superimposed on the warming from increasing greenhouse gases (GHGs).

A third source of uncertainty involves limitations in our current scientific knowledge. Climate models differ in the way they represent various processes (for example, cloud properties, ocean circulation, and aerosol effects). Additionally, climate sensitivity, or how much the climate will warm with a given increase in GHGs (often a doubling of GHG from preindustrial levels), is still a major source of uncertainty. As a result, different models produce small differences in projections of global average change. Scientists often use multiple models to account for the variability and represent this as a range of projected outcomes.

Finally, there is always the possibility that there are processes and feedbacks not yet being included in projections of climate in the future. For example, as the Arctic warms, carbon trapped in permafrost may be released into the atmosphere, increasing the initial warming due to human-caused emissions of GHGs, or an ice sheet may collapse, leading to faster than expected sea level rise.

However, for a given future scenario, the amount of future climate change can be specified within plausible bounds, with those bounds determined not only from the differences in how climate responds to a doubling of GHG concentrations among models but also by utilizing information about climate changes in the past (see Hayhoe et al. 2017[18] for more details).

Key Uncertainties in Temperature Projections

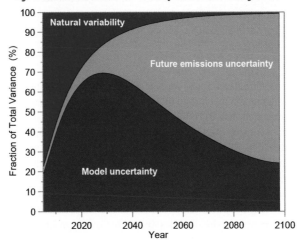

Figure A5.13: The graph shows the change in the fraction of total variance (uncertainty) of three components of total uncertainty in decadal average surface air temperature projections for the contiguous United States. Green represents natural variability, orange represents future emissions uncertainty, and blue represents model or scientific uncertainty (including in climate sensitivity). As the time period becomes more distant, the impact of natural variability becomes less significant due to the smaller variability over a larger period. Future emissions uncertainty increases as time progresses, since we are unable to determine the exact choices that will be made by humans in the future. The influence of model uncertainty on the total uncertainty of how climate will change decreases as the century progresses, due to advances in science and the creation of more accurate and precise assessment systems. This figure shows total uncertainty for the lower 48 states—as the size of the region is reduced, the relative importance of natural variability increases. It is important to note that this figure shows the fractional sources of uncertainty. The total amount of uncertainty increases through time. Source: adapted from Hawkins and Sutton 2009.[19] ©American Meteorological Society. Used with permission.

Is it getting warmer everywhere at the same rate?

Our world is warming overall, but temperatures are not increasing at the same rate everywhere. The average global temperature is projected to continue increasing throughout the remainder of this century due to greenhouse gas (GHG) emissions from human activities. Generally, high latitudes are expected to continue warming more than lower latitudes; coastal and island regions are expected to warm less than interior continent regions.

Temperature changes at a given location are a function of multiple factors, including global and local forces, and both human and natural influences. Though Earth's average temperature is rising, some locations could be cooling due to local factors. In some places, including the U.S. Southeast, temperatures do not show a warming trend over the last century as a whole, although they have been increasing since the 1960s (Ch. 19: Southeast). Possible causes of the observed lack of warming in the Southeast during the 20th century include increased cloud cover and precipitation, increases in the presence of fine particles (called aerosols) in the atmosphere, expanding forests, decreases in the amount of heat conducted from land due to increases in irrigation, and multidecadal variability in sea surface temperatures in both the North Atlantic and the tropical Pacific Oceans. At smaller geographic scales and time intervals, the relative influence of natural variations in climate compared to the human contribution is larger than at the global scale. A lack of warming or a decrease in temperature at an individual location does not negate the fact that, overall, the planet is warming.

Alaska, in contrast to the U.S. Southeast, has been warming twice as fast as the global average since the middle of the 20th century (Ch. 26: Alaska). Statewide average temperatures for 2014–2016 were notably warmer as compared to the last few decades, with 2016 being the warmest on record. Daily record high temperatures in the contiguous United States are now occurring twice as often as record low temperatures. In Alaska, starting in the 1990s, record high temperatures occurred three times as often as record lows, and in 2015, an astounding nine times as often (Ch. 26: Alaska).

Because Earth's climate system still has more energy entering than leaving, global warming has not yet equilibrated to the load of increased GHGs that have already accumulated in the atmosphere (for example, the oceans are still warming over many layers from surface to depth). Some GHGs have long lifetimes (for example, carbon dioxide can reside in the atmosphere for a century or more). Thus, even if the emissions of GHGs were to be sharply curtailed to bring them back to natural levels, it is estimated that Earth is committed to continued warming of more than 1°F by 2100.

At the global scale, some future years will be cooler than the preceding year; some decades could even be cooler than the preceding decade (Figure A5.14). Brief periods of faster temperature increases and also temporary decreases in global temperature can be expected to continue into the future as a result of natural variability and other factors. Nonetheless, each successive decade in the last 30 years has been the warmest in the period of reliable instrumental records (going back to 1850; Figure A5.15). In fact, the rate of warming has accelerated in the past several decades, and 17 of the 18 warmest years have occurred since 2001 (see FAQ "What do scientists mean by the 'warmest year on record'?"). Based on this historical record and assessed scenarios for the future, it is expected that future global temperatures, averaged over climate timescales of 30 years or more, will be higher than preceding periods as a result of emissions of CO_2 and other GHGs from human activities (Ch 2: Climate).

Temperature Change Varies by Region

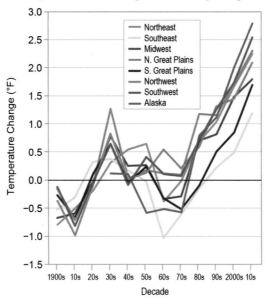

Figure A5.14: This graph shows changes in decadal-averaged temperature relative to the 1901–1960 average for eight of the ten NCA regions (see Front Matter, Figure 1). This figure shows how regional temperatures can be quite variable from decade to decade. All regions, however, have experienced warming over the last three decades or more. The most recent decade, the 2010s, refers to the 6-year period of 2001–2016. Source: adapted from Walsh et al. 2014.[6] Comparable data is not currently available for the Hawai'i and U.S-Affiliated Pacific Islands or U.S. Caribbean regions.

Average Global Temperature Is Increasing

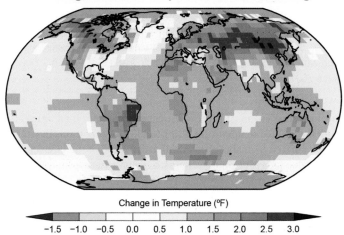

Figure A5.15: This map shows the observed changes in temperature for the 1986 to 2015 period relative to the 1901–1960 average. Shades of red indicate warming, while shades of blue indicate cooling. There are insufficient data in the Arctic Ocean and Antarctica for computing long-term changes. There are substantial regional variations in trends across the planet, though the overall trend is warming. Source: Vose et al. 2012.[20]

What do scientists mean by the "warmest year on record"?

When scientists declare it the "warmest year on record," they mean it's the warmest year since modern global surface temperature record keeping began in 1880. Global temperature data from NASA show that 2016 marked the sixth time this century that a new record high annual average temperature was set (along with 2002, 2005, 2010, 2014, and 2015) and that 17 of the 18 warmest years have occurred since 2001.

The "warmest year on record" means it is the warmest year in more than 130 years of modern record keeping of global surface temperature. Prior to 1880, observations did not cover a large enough area of Earth's surface to enable an accurate calculation of the global average temperature. To calculate the value in recent times, scientists evaluate data from roughly 6,300 stations around the world, on land, ships, and buoys.

The year the last National Climate Assessment was published, 2014, was the warmest year on record at the time, but it was surpassed by 2015, which was then surpassed by 2016. Data from NASA shows that 17 of the 18 warmest years have occurred since 2001, and the 6 warmest years on record have occurred this century (Figure A5.16). However, the global surface temperature is affected by natural variability in addition to climate change, so it is not expected that each year will set a new temperature record.

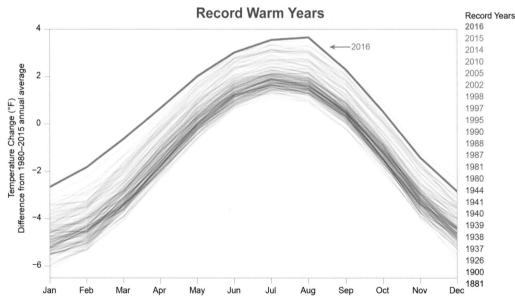

Figure A5.16: This graph shows global, monthly averaged temperature, relative to the 1980–2015 average, plotted over annual temperature cycles from 1880–2017. Record-breaking warm years are listed in the column to the right. The colored lines, shading from gray to blue to purple to red, indicate the years from 1880 to 2017, with 2016, bolded in red, being the hottest year on record. An animation of the complete time series is available online at https://nca2018.globalchange.gov/chapter/appendix-5/#fig-a5-16. Source: NASA.

How do climate projections differ from weather predictions?

The range of possible weather conditions at a specific location on any given day can vary considerably. The climate varies far less for that same location, because it is a measure of weather conditions averaged over 30 years or more. Because the range of possible climate conditions at a given location is much smaller than the range of possible weather conditions, scientists are able to project climate conditions decades into the future.

Projecting how climate may change decades in the future is a different scientific issue than forecasting weather a few days from now. Weather prediction means determining the exact location, time, and magnitude of specific events. Because the range of possible weather conditions can vary so widely, the weather forecast is extremely sensitive to even the smallest uncertainties or errors in our description of the state of the atmosphere at the start of a forecast. The impact of those uncertainties magnifies over time, which makes it very difficult to predict specific weather events at a given location more than a week or two into the future.

Because climate is the average weather at a given location over long periods of time (three decades or more), the range of possible climate conditions at a given location is much smaller than the range of possible weather conditions. For example, the daytime high temperature at a given location may vary by 30°F or more over the course of a day, while the annual average temperature over 30 years may vary by no more than a few degrees (Figure A5.17).

We can project how climate may change over time in response to natural forces, such as changes in incoming solar radiation, and in response to human activities, such as increasing the abundance of greenhouse gases (GHGs) or decreasing particle pollution. These projections are usually expressed in terms of probabilities describing a range of possible outcomes, not in the sort of exact (deterministic) language of many weather forecasts.

The difference between predicting weather and projecting climate is sometimes illustrated with a public health analogy. While it is impossible for us to determine the exact date and time when a particular individual will die, we can easily calculate the average age of death of all Americans for a time period in the past. In this case, weather is like the individual, while climate is like the average. To extend this analogy into the realm of climate change, we can also calculate the average life expectancy of Americans who smoke. We can predict that, on average, smokers will not live as long as nonsmokers. Similarly, we can project what the climate will be like if we emit lower levels of GHGs and what it will be like if we emit more.

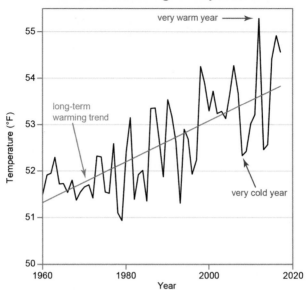

Figure A5.17: This figure shows the annual average surface temperature for the contiguous U.S. (black line) from 1960 to 2017, and the long-term warming trend (red line). Climate change refers to the changes in average weather conditions that persist for an extended period of time, over multiple decades or even longer. Year-to-year and even decade-to-decade, conditions do not necessarily tell us much about long-term changes in climate. One cold year, or even a few cold years in a row, does not contradict a long-term warming trend, just as one hot year does not prove it. Source: adapted from Walsh et al. 2014.[6]

Climate, Weather, and Extreme Events

Was there a "hiatus" in global warming?

Temperature records show that the long-term (30 years or longer) trend in increasing surface temperatures has not ceased. The rate of warming has been faster during some decades and slower during others, but these relatively short periods of time are not the basis for scientists' conclusion that sustained global warming is occurring.

"Global warming" refers to the increase in global average surface temperature that has been observed for more than a century. This warming is clearly revealed in both the surface temperature record and in satellite measurements of lower-atmospheric (troposphere) temperature. While the long-term trend shows warming, scientists expect that the rate of warming will vary from year to year or decade to decade due to the variability inherent in the climate system, or due to short-term changes in climate forcings, such as aerosols (dust, pollution, or volcanic particles) or incoming solar energy (Figure A5.18).

Temporary slowdowns in the rate of warming have occurred earlier in the historical record, even as carbon dioxide concentrations continued to rise. Temporary speedups have also occurred, most notably from the early 1900s to the 1940s and from the 1970s to the late 1990s. Computer simulations of both historical and future climate produce similar variations in the rate of warming, making recent variations in short-term temperature trends unsurprising.

From the mid-1940s to the mid-1970s, there was almost no increase in global temperature, possibly related to an increase in volcanic activity and/or human-caused aerosol emissions. Most notably, for the 15 years following the 1997–1998 El Niño event, the observed rate of temperature increase was smaller than what was projected by some climate models. However, during this period other indicators of climate change continued previous trends associated with warming, such as increasing ocean heat content and decreasing arctic sea ice extent (Figure A5.19; see Wuebbles et al. 2017,[4] Box 1.1).

Short-Term Variability Versus Long-Term Trend

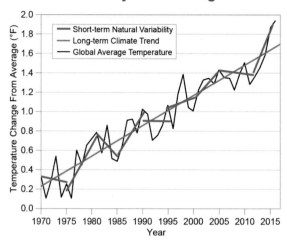

Figure A5.18: Short-term trends in global temperature (blue lines show approximate temperature trends at five-year intervals) can range from decreases to sharp increases. The evidence of climate change is based on long-term trends over 30 years or more (red line). The black line shows the annual average change in global surface temperature from 1970 to 2016 relative to 1901–1960. Source: adapted from Walsh et al. 2014.[6]

Speedups and Slowdowns in Warming

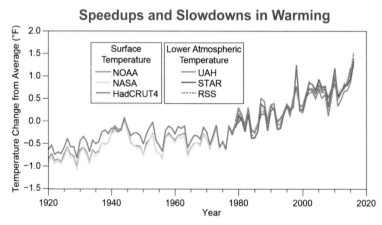

Figure A5.19: The figure shows global annual average surface temperatures (datasets are from NOAA [orange], NASA [yellow], and the United Kingdom's Met Office/University of East Anglia [HadCRUT4, brown]) and lower-atmospheric (tropospheric) temperatures (datasets are from University of Alabama–Huntsville [purple], NOAA [blue], and Remote Sensing Systems [blue dashed]) as compared to 1900–1960 averages. Decades of relatively faster or slower warming are observed within the long-term warming trend. Source: adapted from Trenberth 2015.[21]

What is an extreme event?

An extreme event is a weather or climate-related event that is particularly rare for a given time of year and location. These events include drought, wildfires, floods, severe storms (including hurricanes), heat waves, cold snaps, and heavy rains, and they can have devastating impacts on local communities, infrastructure, the economy, and the environment.

Scientists determine if an event is extreme or not by comparing measurements of weather and climate variables (rainfall, wind speed, temperature, etc.) with thresholds. Events above or below these thresholds are considered rare occurrences, such as events that rank in the highest or lowest 5% of observed values. Several thresholds may be used to define if a single event is considered extreme, and the threshold may change depending on the period of interest (day, month, season, year, etc.) and the chosen reference period (for example, 1961–1990 versus 1900–2000).

It is possible for a single event to meet the definition of an extreme event but not have a large impact. Conversely, it is possible for several types of events that may not be considered extreme individually to cause catastrophic impacts when taken together, such as a sequence of hot days that occur during dry conditions that worsen a drought, or several rainfall events occurring one after another that produce flooding (see Wuebbles et al. 2017, Knutson et al 2017, and Kossin et al. 2017 for more detail on extreme events[4,14,22]).

Have there been changes in extreme weather events?

Yes. Climate change can and has altered the frequency, intensity, duration, or timing of certain types of extreme weather events when compared to past time periods. The harmful effects of severe weather raise concerns about how climate change might alter the risk of such events.

While there have always been extreme events due to natural causes, the frequency and severity of some types of events have increased due to climate change (Figure A5.20) (see also Ch. 2: Climate). As average temperatures have warmed due to emissions of greenhouse gases (GHGs) from human activities, extreme high temperatures have become more frequent and extreme cold temperatures less frequent. From 2001 to 2012, more than twice as many daily high temperature records, as compared to low temperature records, were broken in the United States. With continued increases in the level of GHGs in the atmosphere, the chances for extreme high temperature will continue to increase, with the occurrence of extreme low temperatures becoming less common. Even with much warmer average temperatures later in the century, there may still be occasional record cold snaps, though occurrences of record heat will be more common.

Because warmer air can hold more moisture, heavy rainfall events have become more frequent and severe in some areas and are projected to increase in frequency and severity as the world continues to warm. Both the intensity and rainfall rates of Atlantic hurricanes are projected to increase (see, for example, Ch. 2: Climate, Box 2.5), with the strongest storms getting stronger in a warming climate. Recent research has shown how global warming can alter atmospheric circulation and weather patterns such as the jet stream, affecting the location, frequency, and duration of these and other extremes.[13]

More research would be required to improve scientific understanding of how human-caused climate change will affect other types of extreme weather events important to the United States, such as tornadoes and severe thunderstorms. These events occur over much smaller scales of time and space, which makes observations and modeling more challenging. Projecting the future influence of climate change on these events can also be complicated by the fact that some of the risk factors for these events may increase while others may decrease.[2,4,22]

Extreme Temperature and Precipitation Events

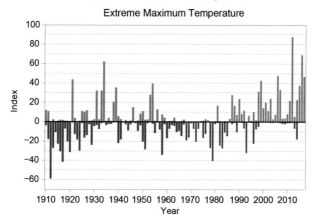

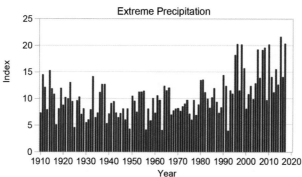

Figure A5.20: The top panel shows the percentage of land area in the contiguous United States that experienced maximum temperatures greatly above or below normal (upper or lower 10th percentile, respectively). The bottom panel shows the percentage of the land area for the contiguous United States that experienced extreme 1-day precipitation amounts that were greatly above normal. In the past 25 years, a much greater area of the country has experienced warmer extreme maximum temperatures and extreme rainfall. Sources: NOAA NCEI and CICS-NC.

Can specific weather or climate-related events be attributed to climate change?

While it is difficult to attribute a specific weather or climate-related event to any one cause, climate change can affect whether an event was more or less likely to occur. Climate change can also influence the severity of these events. Our ability to detect the influence of human-caused warming on particular kinds of extreme events depends both on the length and quality of our historical records of those events, as well as how well we can simulate the environmental processes that produce and sustain them.

Extreme event attribution is a relatively recent scientific advancement that seeks to determine whether climate change altered the likelihood of occurrence of a given extreme event.[14,23] A long-term, high-quality record of a given type of event and a computer model capable of producing a realistic simulation of the event are needed in order to assess the influence of climate change. Because of these data and modeling constraints, our ability to detect the influence of human-caused global warming on heat waves and, to a lesser extent, heavy rainfall events is better at present than our ability to detect its influence on tornadoes or hurricanes. As scientists collect more data and develop more advanced tools, they will be able to better quantify cause-and-effect relationships in the climate system, which should improve their ability to attribute how much human-caused climate change contributes to specific weather and climate-related events.

One example of event attribution comes from the recent California drought, where scientists found that human-caused climate change contributed 8%–27% to the severity of the drought.[24] Droughts are frequent in the Southwest and occur regardless of human activity, but human-caused climate change leads to increased evaporation and decreased soil moisture, intensifying droughts during periods of little rain.[14]

Could climate change make Atlantic hurricanes worse?

Atlantic hurricane activity has increased since the 1970s, but the relatively short length of high-quality hurricane records does not yet allow us to say how much of that increase is natural and how much may be due to human activity. With future warming, hurricane rainfall rates are likely to increase, as will the number of very intense hurricanes, according to both theory and numerical models. However, models disagree about whether the total number of Atlantic hurricanes will increase or decrease. Rising sea level will increase the threat of storm surge flooding during hurricanes.

Hurricane activity is undeniably linked to sea surface temperatures (see Ch. 2: Climate, Box 2.5 for a discussion on the 2017 Atlantic hurricane season). Other influences being equal, warmer waters yield stronger hurricanes with heavier rainfall. The tropical Atlantic Ocean has warmed over the past century, at least partly due to human-caused emissions of greenhouse gases. However, high-quality records of Atlantic hurricanes are too short to reliably separate any long-term trends in hurricane frequency, intensity, storm surge, or rainfall rates from natural variability.[22] This does not mean that no trends exist, only that the data record is not long enough to determine the cause.

Most models agree that climate change through the 21st century is likely to increase the average intensity and rainfall rates of hurricanes in the Atlantic and other basins. Models are less certain about whether the average number of storms per season will increase or decrease. Early modeling raised the possibility of a significant future increase in the number of Category 4 and 5 storms in the Atlantic (Figure A5.21). While that remains possible, the most recent high-resolution modeling provides mixed messages: some models project increases in the number of the basin's strongest storms, and others project decreases.[22]

Regardless of any human-influenced changes in storm frequency or intensity, rising sea level will increase the threat of storm surge flooding during hurricanes (Ch. 8: Coastal; Ch. 18: Northeast; Ch. 19: Southeast; Ch. 20: U.S. Caribbean; Ch. 23: S. Great Plains).

Category 4 and 5 Hurricane Formation: Now and in the Future

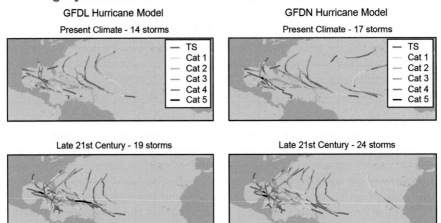

Figure A5.21: These maps show computer-simulated tracks and intensities of hurricanes reaching Categories 4 and 5 (intensity based on wind speeds ranging from TS for tropical storm strength up to Category 1 through Category 5 hurricanes). The top panels show hurricane tracks from two different models under current climate conditions (1980–2006). The bottom panels show projections from the same models but for late-21st century (2081–2100) conditions, both under the lower scenario (RCP4.5). These projections show an increase in the frequency of Category 4 and 5 hurricanes, with a higher tendency of these storms to shift towards the Gulf of Mexico, Florida, and the Caribbean (as opposed to remaining in the open Atlantic Ocean). Source: adapted from Knutson et al. 2013.[25] ©American Meteorological Society. Used with permission.

Societal Effects

How is climate change affecting society?

Climate change is altering the world around us in ways that become increasingly evident with each passing decade. Natural and human systems that we rely on are being impacted by more intense precipitation events, rising sea level, and a warming ocean and will be impacted by projected increases in the frequency of droughts and heat waves and other extreme weather patterns.

Many people are already being affected by the changes that are occurring, and more will be affected as these changes continue to unfold (Figure A5.22). In the Northeast and Northwest, fishing communities have to adapt to increasing ocean temperatures and acidification that impact fish and shellfish (Ch. 9: Oceans; Ch. 18: Northeast; Ch. 24: Northwest). Coastal communities, especially those located on islands, will need to confront rising sea levels, which are already contaminating freshwater supplies, flooding streets during high tides, and exacerbating storm surge flooding (Ch. 8: Coastal; Ch. 19: Southeast; Ch. 20: U.S. Caribbean; Ch. 27: Hawai'i and Pacific Islands). Shifts in the timing of the seasons and changes in the location of plants and animals affect communities dependent on those resources for tourism, economy, and/or cultural purposes (Ch. 7: Ecosystems; Ch. 15: Tribes; Ch. 26: Alaska).

Changes are not only happening in the oceans and along the coast. Farmers, the livestock they tend, and other outdoor laborers are expected to be adversely affected by warmer temperatures, an increasing frequency of heat waves, and an increasing number of warm nights (Ch. 10: Ag & Rural; Ch. 14: Human Health; Ch. 19: Southeast; Ch. 23: S. Great Plains). Some communities may have to adapt to both an increase in the frequency of drought and more rain falling as heavy precipitation, while deteriorating water infrastructure compounds those risks (Ch. 3: Water; Ch. 17: Complex Systems; Ch. 22: N. Great Plains; Ch. 25 Southwest). The geographic range and distribution of some pests and pathogens are projected to change in some regions, exposing livestock and crops to new or additional stressors and exposing more people to diseases transmitted by those pests (Ch. 14: Human Health; Ch. 21: Midwest).

Infrastructure across the country, which supports economic activity, is increasingly being tested and impacted by climate change, including airport runways affected by increased surface temperature and coastal streets inundated by high tide flooding (Ch. 12: Transportation). Much of the current built environment throughout the country has been developed based on the assumption that future climate will be similar to that of the past, which is no longer a valid assumption (Ch. 11: Urban). In general, the larger and faster the changes in climate, the more difficult it is for human and natural systems to adapt. Adaptation efforts not only help communities become more resilient, they may also create new jobs and help stimulate local economies (see FAQ "What are climate change mitigation, adaptation, and resilience?").

Americans Respond to the Impacts of Climate Change

Alaska

Impact	Action
The physical and mental health of rural Alaskans is increasingly challenged by unpredictable weather and other environmental changes.	The Alaska Native Tribal Health Consortium's Center for Climate and Health is using novel adaptation strategies to reduce climate-related risk including difficulty in harvesting local foods and more hazardous travel conditions.

Northern Great Plains

Impact	Action
Flash droughts and extreme heat illustrate sustainability challenges for ranching operations, with emergent impacts on rural prosperity and mental health.	The National Drought Mitigation Center is helping ranchers plan to reduce drought and heat risks to their operations.

Midwest

Impact	Action
Increasing heavy rains are leading to more soil erosion and nutrient loss on Midwestern cropland.	Iowa State developed a program using prairie strips in farm fields to reduce soil and nutrient loss while increasing biodiversity.

Northwest

Impact
Wildfire increases and associated smoke are affecting human health, water resources, timber production, fish and wildlife and recreation.

Action
Federal forests have developed adaptation strategies for climate change that include methods to address increasing wildfire risks.

Northeast

Impact
Water, energy, and transportation infrastructure are affected by snow storms, drought, heat waves, and flooding.

Action
Cities and states throughout the region are assessing their vulnerability to climate change and making investments to increase infrastructure resilience.

Southwest

Impact	Action
Drought in the Colorado River basin reduced Lake Mead by over half since 2000, increasing risk of water shortages for cities, farms, and ecosystems.	Seven U.S. state governments and U.S. and Mexico federal governments mobilized users to converse water, keeping the lake above a critical level.

Southern Great Plains

Impact	Action
Hurricane Harvey's landfall on the Texas coast in 2017 was one of the costliest natural disasters in U.S. history.	The Governor's Commission to Rebuild Texas was created to support the economic recovery and rebuilding of infrastructure in affected Texas communities.

Southeast

Impact	Action
Flooding in Louisiana is increasing from extreme rainfall.	The Acadiana Planning Commission in Louisiana is pooling hazard reduction funds to address increasing flood risk.

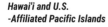

Hawai'i and U.S. -Affiliated Pacific Islands

Impact	Action
The 2015 coral bleaching event resulted in an average mortality of 50% of the coral cover in western Hawai'i alone.	A state working group generated management options to promote recovery and reduce threats to coral reefs.

U.S. Caribbean

Impact	Action
Damages from the 2017 hurricanes have been compounded by slow recovery of energy, communications, and transportation systems, impacting all social and economic sectors.	The U.S. Virgin Islands Governor's Office led a workshop aimed at gathering lessons from the initial hurricane response and establishing a framework for recovery and resilience.

Figure A5.22: This map shows climate-related impacts that have occurred in each region since the Third National Climate Assessment in 2014 and response actions that are helping the region address related risks and costs. These examples are illustrative; they are not indicative of which impact is most significant in each region or which response action might be most effective. Source: NCA4 Regional Chapters.

What is the social cost of carbon?

The social cost of carbon is an estimate of the monetary value of the cumulative damages caused by long-term climate change due to an additional amount of carbon dioxide (CO_2) emitted. This value quantifies the potential benefits of a reduction in CO_2 emissions.

The social cost of carbon (SCC) includes the economic costs of climate change that will be felt in market sectors such as agriculture, energy services, and coastal resources, as well as nonmarket impacts on human health and ecosystems, to name a few.[26] SCC values are computed by simulating the "causal chain" from greenhouse gas emissions to physical climate change to climate damages in order to estimate the additional damages over time incurred from an additional metric ton of CO_2.[27] This value can be used to inform climate risk management decisions at national, state, and corporate levels, as well as in regulatory impact analysis to evaluate benefits of marginal CO_2 reductions—for example, in rules affecting appliance efficiency, power generation, industry, and transportation, such as the benefits of increased vehicle gas mileage standards. As with many complex, interacting systems, it is challenging to develop comprehensive SCC estimates, but this is an active area of research guided by recent recommendations from the National Academies of Sciences, Engineering, and Medicine to keep up with the current state of scientific knowledge, better characterize key uncertainties, and improve transparency.[28] Notably, estimating the SCC depends on normative social values such as time preference, risk aversion, and equity considerations that can lead to a range of values. Ongoing interdisciplinary collaborations and research findings from the climate change impacts, adaptation, and vulnerability literature—including those discussed in the Fourth National Climate Assessment—are being used to improve the robustness of climate damage quantification and, thus, SCC estimates.

What are climate change mitigation, adaptation, and resilience?

"Mitigation," "adaptation," and "resilience" are related but different terms in the context of climate change. Mitigation refers to actions that reduce the amount and speed of future climate change by reducing emissions of greenhouse gases (GHGs)or removing carbon dioxide from the atmosphere. Adaptation refers to adjustments in natural or human systems in response to a new or changing environment that exploit beneficial opportunities or moderate negative effects. Thus, adaptation is closely related to resilience, which is the capacity to prevent, withstand, respond to, and recover from a disruption with minimum damage to social well-being, the economy, and the environment.

Mitigation efforts can reduce emissions or increase storage of GHGs. For example, shifting from fossil fuels to low-carbon energy sources will generally result in the reduction of GHG emissions into the atmosphere. Mass transit, energy-efficient buildings, and electric vehicles can be used instead of high-emission alternatives. Land-use changes that increase the amount of carbon stored in soil and biomass, as well as some geoengineering techniques, constitute mitigation efforts that take carbon dioxide (CO_2) out of the atmosphere (see FAQ "Can geoengineering be used to remove carbon dioxide from the atmosphere or otherwise reverse global warming?") (see also Ch. 29: Mitigation).

Adaptation involves policies, strategies, and technologies designed to reduce the risk of harm from climate-related impacts. Some adaptation actions are technical engineering solutions designed to address specific impacts, such as building a seawall in the face of sea level rise or breeding new crops that do well in the context of drought. Other adaptation actions involve decision-making processes, policies, or approaches that bring people together to support coordinated action (Ch. 28: Adaptation). Adaptation often involves incremental adjustments to current systems, but larger transformations may be necessary, especially as some systems cross thresholds or tipping points.

Adaptation and mitigation actions can be undertaken simultaneously to reduce concentrations of GHGs in the atmosphere while also reducing the risk of climate-related impacts. Both adaptation and mitigation can have co-benefits—societal benefits that are not necessarily related to climate change (Ch. 29: Mitigation). For example, a new coastal restoration project to plant a mangrove forest will remove CO_2 from the atmosphere while providing valuable ecosystem services—a buffer against storm surges, reduced erosion, habitat for wildlife, and filtration of human pollutants (Ch. 8: Coastal).

Climate resilience refers to the capacity of a human or natural system to respond to and recover from climate-related hazards, such as droughts or floods, in ways that maintain their essential or valued identity, functions, and structure. Resilient systems respond to climate stressors or impacts with less harm while also improving their ability to absorb future impacts and maintaining capacity for adaptation and learning. A resilient rural community might have the capacity to share knowledge and resources to help farmers deal with droughts while improving their ability to absorb future impacts by building long-term structures to conserve water resources (Ch. 24: Northwest). Resilience can be bolstered by diversity (such as species diversity or employment diversity), redundancy (the ability for one part of the system to take over essential functions if another is damaged), social networks, knowledge sharing, and good governance (Ch. 7: Ecosystems).

Is timing important for climate mitigation?

Yes. The choices made today largely determine what impacts may occur in the future. Carbon dioxide can persist in the atmosphere for a century or more, so emissions released now will still be affecting climate for years to come. The sooner greenhouse gas (GHGs) emissions are reduced, the easier it may be to limit the long-term costs and damages due to climate change. Waiting to begin reducing emissions is likely to increase the damages from climate-related extreme events (such as heat waves, droughts, wildfires, flash floods, and stronger storm surges due to higher sea levels and more powerful hurricanes).

The effect of increasing atmospheric concentrations of carbon dioxide (CO_2) and other GHGs on the climate system can take decades to be fully realized. The resulting change in climate and the impacts of those changes can then persist for centuries. The longer these changes in climate continue, the greater the resulting impacts; some systems may not be able to adapt if the change is too much or too fast.

The long-term equilibrium temperature from GHG emissions will be a function of cumulative emissions over time, not the specific year-to-year emissions. Thus, staying within a specific warming target will depend on the total net emissions (including increases in carbon uptake) over a given future period.

However, the timing and nature of changes are important in both reducing short-term warming and meeting any particular long-term warming limit. Long-term reductions in the rate and magnitude of global warming can be made by reducing total emissions of CO_2. Near-term reductions in the rate of climate change can be made by reducing human-caused emissions of short-lived but highly potent GHGs such as methane and hydrofluorocarbons. These pollutants remain in the atmosphere from weeks to about a decade—much shorter than CO_2—but have a much greater warming influence than CO_2 (Figure A5.23).[17]

Benefit of Earlier Action to Reduce Emissions

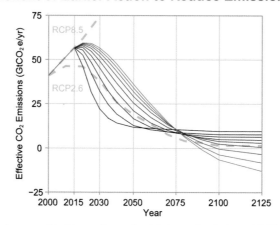

Figure A5.23: This figure shows possible future pathways for global annual emissions of GHGs for which the global mean temperature would likely (66%) not exceed 3.6°F (2°C) above the preindustrial average. The black curves on the bottom show the fastest reduction in emissions, with rapid near-term mitigation and little to no negative emissions required in the future. The red curves on top show slower rates of mitigation, with slow near-term reductions in emissions and large negative emission requirements in the future. Here, the annual global GHG emissions are in units of gigatons of CO_2 equivalent, a measurement that expresses the warming impact of all GHGs in terms of the equivalent amount of CO_2. Source: adapted from Sanderson et al. 2016.[29]

Are there benefits to climate change?

While some climate changes currently have beneficial effects for specific sectors or regions, many studies have concluded that climate change will generally bring more negative effects than positive ones in the future. For example, current benefits of warming include longer growing seasons for agriculture, more carbon dioxide for plants, and longer ice-free periods for shipping on the Great Lakes. However, longer growing seasons, along with higher temperatures and increased carbon dioxide levels, can increase pollen production, intensifying and lengthening the allergy season. Longer ice-free periods on the Great Lakes can result in more lake-effect snowfalls.

Many analyses of this question have concluded that climate change will, on balance, bring more negative effects than positive ones in the future. This is largely because our society and infrastructure have been built for the climate of the past, and changes from those historical climate conditions impose costs and management challenges (Ch. 11: Urban). For example, while longer warm seasons may provide a temporary economic boon to coastal communities reliant on tourism, many of these same areas are vulnerable not only to sea level rise but also to risks from ocean acidification and warmer waters that can impact the ecosystems (such as coral reefs) that bring people to the coasts (Ch. 8: Coastal). As another example, while some studies have shown that certain crops in certain regions may benefit from additional carbon dioxide (CO_2) in the atmosphere (sometimes referred to as the CO_2 fertilization effect), these potential gains are expected to be offset by crop stress caused by higher temperatures, worsening air quality, and strained water availability (see FAQ "How do higher carbon dioxide concentrations affect plant communities and crops?") (see also Ch. 10: Ag & Rural). Furthermore, any accrued benefits are likely to be short-lived and depreciate significantly as warming continues through the century and beyond.

Are some people more vulnerable to climate change than others?

Yes. Climate change affects certain people and populations differently than others. Some communities have higher exposure and sensitivity to climate-related hazards than others. Some communities have more resources to prepare for and respond to rapid change than others. Communities that have fewer resources, are underrepresented in government, live in or near deteriorating infrastructure (such as damaged levees), or lack financial safety nets are all more vulnerable to the impacts of climate change.

Vulnerability here refers to the degree to which physical, biological, and socioeconomic systems are susceptible to and unable to cope with adverse impacts of climate change. Vulnerability encompasses sensitivity, adaptive capacity, exposure, and potential impacts. For example, older people living in cities with no air conditioning have less adaptive capacity and increased sensitivity and vulnerability to heat stress during extreme heat events (Ch. 14: Human Health). Communities that live on atolls in the Marshall Islands have high exposure and are acutely at risk to sea level rise and saltwater intrusion due to the low land height and small land area (Ch. 27: Hawai'i & Pacific Islands). A history of neglect, political or otherwise, in a given neighborhood can result in dilapidated infrastructure, which in turn can lead to situations such as levee failures, making whole communities vulnerable to flooding and other potential impacts (Ch. 14: Human Health). Poverty can make evacuation during storm events challenging and can make rebuilding or relocating harder following an extreme event. In some Indigenous communities, lack of water and sanitation systems can put people at risk during drought (Ch. 15: Tribes). Additionally, some subpopulations are already more affected by environmental exposures, such as air pollution or extreme heat. If communities or individuals experience a combination of these vulnerability factors, they are at even greater risk. Vulnerable communities and individuals face these disparities today and will likely face increased challenges in the future under a changing climate.

How will climate change impact economic productivity?

Many impacts of climate change are expected to have negative effects on economic productivity, such as increased prices of goods and services. For example, increased exposure to extreme heat may reduce the hours some individuals are able to work. Physical capital—such as food, equipment, and property—that is derived from the production of goods and services may be impacted because of lower production and higher costs as a result of climate change. Sea level rise, stronger storm surges, and increased heavy downpours that cause flooding can disrupt supply chains or damage properties, structures, and infrastructure that form the backbone of the Nation's economy.

High temperatures and storm intensity, which are both linked to more deaths and illness, are projected to increase due to climate change, which would in turn increase health care costs for medical treatment. At the same time, these health effects directly impact labor markets. Workers in industries with the greatest exposure to weather extremes may decrease the amount of time they spend at work, while workers across a wide range of sectors may find their productivity impaired while on the job (Ch. 14: Human Health). These labor market impacts translate into lower earnings for workers and firms.[30,31]

Climate change is likely to affect physical capital that serves as an important input to economic production. In farming, where weather is a key determinant of agricultural yield, increasing temperatures and drought may lead to net decreases in the amount of food that farms produce (Ch.10: Ag & Rural).[32] Extreme heat can also cause manufacturing equipment to break down with greater frequency, while rising sea levels and increased storm intensity can destroy equipment and property across all types of economic activities along American coastlines.[30,33]

In addition to damaging private property, increased weather extremes can destroy vital public infrastructure, such as roads, bridges, and ports. Since this infrastructure is an integral part of supply chains that drive the American economy, a disruption in their accessibility—or even their destruction—can have large impacts on corporate profits, while their repairs require a diversion of resources away from other useful government projects or an increase in taxes to finance reconstruction (Ch. 11: Urban).[34,35]

Can we slow climate change?

Yes. While we cannot stop climate change overnight, or even over the next several decades, we can limit the amount of climate change by reducing human-caused emissions of greenhouse gases (GHGs). Even if all human-related emissions of carbon dioxide and other GHGs were to stop today, Earth's temperature would continue to rise for a number of decades and then slowly begin to decline. Ultimately, warming could be reversed by reducing the amount of GHGs in the atmosphere. The challenge in slowing or reversing climate change is finding a way to make these changes on a global scale that is technically, economically, socially, and politically viable.

The most direct way to significantly reduce the magnitude of future climate change is to reduce the global emissions of GHGs. Emissions can be reduced in many ways, and increasing the efficiency of energy use is an important component of many potential strategies (Ch. 29: Mitigation). For example, because the transportation sector accounts for about 29% of the energy used in the United States, developing and driving more efficient vehicles and changing to fuels that do not contribute significantly to GHG emissions over their lifetimes would result in fewer emissions per mile driven. A large amount of energy in the United States is also used to heat and cool buildings, so changes in building design could dramatically reduce energy use (Ch 29: Mitigation). While there is no single approach that will solve all the challenges posed by climate change, there are many options that can reduce emissions and help prevent some of the potentially serious impacts of climate change (Figure A5.24).[17]

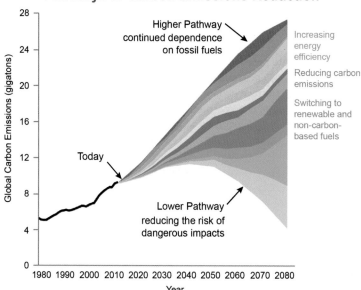

Figure A5.24: Reducing carbon emissions from a higher scenario (RCP8.5) to a lower scenario (RCP4.5) can be accomplished with a combination of many technologies and policies. In this example, these emissions reduction "wedges" could include increasing the energy efficiency of appliances, vehicles, buildings, electronics, and electricity generation (orange wedges); reducing carbon emissions from fossil fuels by switching to lower-carbon fuels or capturing and storing carbon (blue wedges); and switching to renewable and non-carbon-emitting sources of energy, including solar, wind, wave, biomass, tidal, and geothermal (green wedges). The shapes and sizes of the wedges shown here are illustrative only. Source: adapted from Walsh et al. 2014.[6]

Can geoengineering be used to remove carbon dioxide from the atmosphere or otherwise reverse global warming?

In theory, it may be possible to reverse some aspects of global warming through technological interventions called geoengineering, which can complement mitigation and adaptation. But many questions remain. Geoengineering approaches generally fall under two categories: 1) carbon dioxide removal and 2) reducing the amount of the sun's energy that reaches Earth's surface. Due to uncertain costs and risks of some geoengineering approaches, more traditional mitigation actions to reduce emissions of greenhouse gases (GHGs) are generally viewed as more feasible for avoiding the worst impacts from climate change currently. However, targeted studies to determine the feasibility, costs, risks, and benefits of various geoengineering techniques could help clarify the impacts.

Removal of carbon dioxide (CO_2) from the atmosphere could be undertaken by applying land management methods that increase carbon storage in forests, soils, wetlands, and other terrestrial or aquatic carbon reservoirs. Trees and plants draw down CO_2 from the atmosphere during photosynthesis and store it in plant structures. Reforesting large tracts of deforested lands would help reduce atmospheric concentrations of CO_2. New technologies could also be used to capture CO_2 either directly from the atmosphere or at the point where it is produced (such as at coal-fired power plants) and store it underground. However, CO_2 removal may be costly and has long implementation times, and the removal of CO_2 from the atmosphere must be essentially permanent if climate impacts are to be avoided.[17,36]

Solar radiation management (SRM) is an intentional effort to reduce the amount of sunlight that reaches Earth's surface by increasing the amount of sunlight reflected back to space. Since SRM does not reverse the increased concentrations of CO_2 and other GHGs in the atmosphere, this approach does not address direct impacts from elevated CO_2, such as damage to marine ecosystems from increasing ocean acidification.[17,37] Instead, it introduces another human influence on the climate system that partially cancels some of the effects of increased GHGs in the atmosphere. SRM methods include making clouds brighter and more reflective, injecting reflective aerosol particles into the upper or lower atmosphere, or increasing the reflectivity of Earth's surface. SRM can work in conjunction with CO_2 removal and other mitigation efforts and can be phased out over time. Yet this method would require sustained costs, has not been well studied, and could have harmful unintended consequences, such as stratospheric ozone depletion.[38]

Ecological Effects

What causes global sea level rise, and how will it affect coastal areas in the coming century?

Global sea level is rising, primarily in response to two factors: 1) thermal expansion of ocean waters and 2) melting of land-based ice, both due to climate change. Thermal expansion refers to the physical expansion (or increase in volume) of water as it warms. Melting of mountain glaciers and the Antarctic and Greenland ice sheets contributes additional water to the oceans, thereby raising global average sea level. Global average sea level has risen 7–8 inches since 1880, and about 3 inches of that has occurred since 1993. Sea level rise will increasingly contribute to high tide flooding and intensify coastal erosion over the coming century.

At any given location, the situation is more complicated because other factors come into play. For example, coastlands are rising in some places and sinking in others due to both natural causes (such as tectonic shifts) and human activities (such as groundwater or hydrocarbon extraction). Where coastlands are rising as fast as (or faster than) sea level, relative local sea level may be unchanged (or decreasing). Where coastlands are sinking (called subsidence), relative local sea level may be rising faster than the global average (Figure A5.25) (see also Ch. 23: S. Great Plains). Other variables can influence relative sea level locally, including natural climate variability patterns (for example, El Niño/La Niña events) and regional shifts in wind and ocean current patterns.[39]

Global sea level rise is already affecting the U.S. coast in many locations (Ch. 8: Coastal). High tide flooding with little or no storm effects (also referred to as nuisance, sunny-day, or recurrent flooding), coastal erosion, and beach and wetland loss are all increasingly common due to decades of local relative sea level rise (Ch. 19: Southeast).[39] Sea level is expected to continue rising at an accelerating rate this century under either a lower or higher scenario (RCP4.5 or RCP8.5), increasing the frequency of high tide flooding, intensifying coastal erosion and beach and wetland loss, and causing greater damage to coastal properties and structures due to stronger storm surges (Ch. 18: Northeast; Ch. 8: Coastal). Relative local sea level rise projections can be visualized at https://coast.noaa.gov/digitalcoast/tools/slr.html.

Relative Sea Level Projected to Rise Along Most U.S. Coasts

Lower Scenario (RCP4.5)

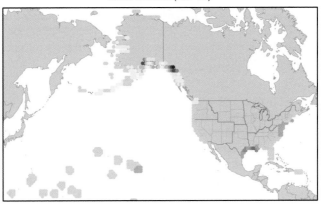

Higher Scenario (RCP8.5)

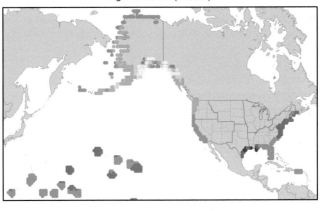

Relative Sea Level Change (feet)

-6   -4   -2   0   2   4   6

Figure A5.25: The maps show projections of change in relative sea level along the U.S. coast by 2100 (as compared to 2000) under the lower and higher scenarios (RCP4.5 and RCP8.5, top and bottom panels, respectively).[39] Globally, sea levels will continue to rise from thermal expansion of the ocean and melting of land-based ice masses (such as Greenland, Antarctica, and mountain glaciers). Regionally, however, the amount of sea level rise will not be the same everywhere. Where land is sinking (as along the Gulf of Mexico coastline), relative sea level rise will be higher, and where land is rising (as in parts of Alaska), relative sea level rise will be lower. Changes in ocean circulation (such as the Gulf Stream) and gravity effects due to land ice melt will also alter the heights of the ocean regionally. Sea levels are expected to continue to rise along almost all U.S. coastlines, and by 2100, under the higher scenario, coastal flood heights that today cause major damages to infrastructure would become common during high tides nationwide. Source: adapted from Sweet et al. 2017.[40]

How does global warming affect arctic sea ice cover?

The Arctic region has warmed by about 3.6°F since 1900—double the rate of the global temperature increase. Consequently, sea ice cover has declined significantly over the last four decades. In the summer and fall, sea ice area has dropped by 40% and sea ice volume has dropped 70% relative to the 1970s and earlier. Decline in sea ice cover plays an important role in arctic ecosystems, ultimately impacting Alaska residents.

Arctic sea ice today is in the most reduced state since satellite measurements began in the late 1970s, and the current rate of sea ice loss is also unprecedented in the observational record (Figures A5.26 and A5.27) (see also Ch. 2: Climate). Arctic sea ice cover is sensitive to climate change because strong self-reinforcing cycles (positive feedbacks) are at play. As sea ice melts, more open ocean is exposed. Open ocean (a dark surface) absorbs much more sunlight than sea ice (a reflective white surface). That extra absorbed sunlight leads to more warming locally, which in turn melts more sea ice, creating a positive feedback (Ch. 2: Climate). Annual average arctic sea ice extent has decreased between 3.5% and 4.1% per decade since the early 1980s, has become thinner by 4.3 to 7.5 feet, and has started melting earlier in the year. September sea ice extent, when the arctic sea ice is at a minimum, has decreased by 10.7% to 15.9% per decade since the 1980s. Scientists project sea ice-free summers in the Arctic by the 2040s (Figure A5.27) (see Ch. 26: Alaska).[2]

Arctic sea ice plays a vital role in arctic ecosystems. Changes in the extent, duration, and thickness of sea ice, along with increasing ocean temperature and ocean acidity, alter the distribution of Alaska fisheries and the location of polar bears and walruses, all of which are important resources for Alaska residents, particularly coastal Native Alaska communities (Ch. 26: Alaska). Winter sea ice may keep forming in a warmer world, but it could be much reduced compared to the present (see Taylor et al. 2017[2] for more details).

Annual Minimum Sea Ice Extent Decreasing

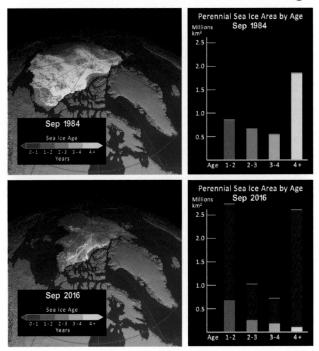

Figure A5.26: Both the extent and the age of the September sea ice cover are shown for 1984 (top) and 2016 (bottom). The colors of the bars on the right panels correspond to the colors used to indicate the age of the sea ice in the panels on the left. The green bars on the graphs on the right mark the maximum extent for each age range during the record. The year 1984 is representative of September sea ice characteristics during the 1980s. Over time, September sea ice extent and the amount of multiyear ice have greatly decreased. The years 1984 and 2016 are selected as endpoints in the timeseries. A movie of the complete time series is available at http://svs.gsfc.nasa.gov/cgi-bin/details.cgi?aid=4489. Source: adapted from NASA 2016.[41]

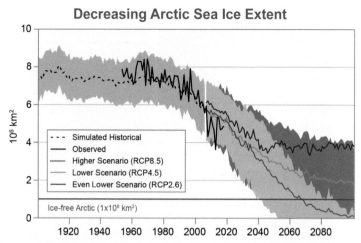

Decreasing Arctic Sea Ice Extent

Figure A5.27: This graph shows historical simulations of arctic sea ice extent starting in 1900 (dotted black line), observations of arctic sea ice extent (solid black line), and future projections of arctic sea ice extent (colored lines) from 2005 through 2100 under three RCP scenarios. The projections shown are the average values from a set of climate model simulations, and the shaded pink and green regions indicate one-standard-deviation confidence intervals around the average values for the higher and lower scenarios, respectively. Source: adapted from Stroeve and Notz 2015.[42] ©2015 Elsevier B.V. All rights reserved.

Is Antarctica losing ice? What about Greenland?

Yes. Overall, the ice sheets on both Greenland and Antarctica, the largest areas of land-based ice on the planet, are losing ice as the atmosphere and oceans warm. This ice loss is important both as evidence that the planet is warming and because it contributes to rising sea levels.

The Antarctic ice sheet is up to three miles deep and contains enough water to raise sea level about 200 feet. Because Antarctica is so cold, there is little melting of the ice sheet, even in summer. However, the ice flows towards the ocean where above-freezing ocean water speeds up the melting process, which breaks the ice into free-floating icebergs (a process called calving). Melting, calving, and the flow of ice into the oceans around Antarctica—especially on the Antarctic Peninsula—have all accelerated in recent decades, and the result is that Antarctica is losing about 100 billion tons of ice per year (contributing about 0.01 inch per year to sea level rise; Figure A5.28).[39] While there has been slight growth in some parts of the Antarctic ice sheet, the gain is more than offset by ice mass loss elsewhere, especially in West Antarctica and along the Antarctic Peninsula. The West Antarctic ice sheet, which contains enough ice to raise global sea level by 10 feet, is likely to lose ice much more quickly if its ice shelves disintegrate. Additionally, warming oceans under the ice sheet are melting the areas where ice sheets go afloat in West Antarctica, exacerbating the risk of more rapid melt in the future.

Greenland contains only about one-tenth as much ice as the Antarctic ice sheet, but if Greenland's ice sheet were to entirely melt, global sea level would still rise about 20 feet. (For additional information on the impacts of sea level rise on the United States directly, see Ch. 8: Coastal; Ch. 18: Northeast; Ch. 19: Southeast; and Ch. 20: U.S. Caribbean.) Annual surface temperatures in Greenland are warmer than Antarctica, so melting occurs over large parts of the surface of Greenland's ice sheet each summer. Greenland's melt area has increased over the past several decades (Figure A5.28). The Greenland ice sheet is presently thinning at the edges (especially in the south) and slowly thickening in the interior, increasing the steepness of the ice sheet, which has sped up the flow of ice into the ocean over the past decade. This trend will likely continue as the surrounding ocean warms. Greenland's ice loss has increased substantially in the past decade, losing ice at an average rate of about 269 billion tons per year from April 2012 to April 2016 (contributing over 0.02 inch per year to sea level rise).[4]

Greenland and Antarctica Are Losing Ice

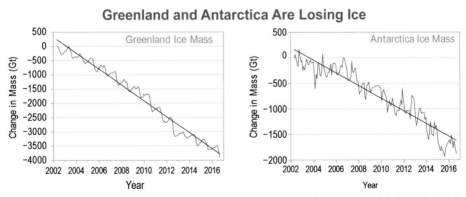

Figure A5.28: The graphs show satellite measurements of the change in ice mass for the two polar ice sheets through August 2016 as compared to April 2002. Both the Greenland and Antarctic ice sheets are losing ice as the atmosphere and oceans warm. Source: adapted from Wouters et al. 2013.[43] Reprinted by permission from Macmillan Publishers Ltd., ©2013.

How does climate change affect mountain glaciers?

Glacier retreat is one of the most important lines of evidence for global warming. Around the world, glaciers in most mountain ranges are receding at unprecedented rates. Many glaciers have disappeared altogether this century, and many more are expected to vanish within a matter of decades. Glaciers will still be around within the next century, but they will be more isolated, closer to the poles, and at higher elevations.

Glaciers are critical freshwater reservoirs that slowly release water over warmer months, which helps sustain freshwater streamflows that provide drinking and irrigation water, as well as hydropower to downstream communities. However, increasing temperatures and decreasing amounts of precipitation falling as snow are major drivers of glacial retreat (see Ch. 2: Climate; Ch. 22: N. Great Plains; Ch. 24: Northwest; Ch. 26: Alaska). Glaciers retreat when melting and evaporation outpace the accumulation of new snow. Slope, altitude, ice flow, location, and volume also contribute to the speed and extent of glacial retreat, which complicates the relationship between increasing temperature and glacial melt. Due to these local factors, not all glaciers globally are retreating. For example, melting may slow as the glaciers retreat to the upper slopes, under headwalls and steep cliffs, and into more shaded areas.

In recent decades, the mountains of Glacier National Park (GNP) in Montana have experienced an increase in summer temperatures and a reduction in the winter snowpack that forms the mountain glaciers. The annual average temperature in GNP has increased by 2.4°F since 1900, spring and summer minimum temperatures have risen, and the percentage of precipitation that comes as rain rather than snow has increased.[44,45,46] Mountain snowpacks now hold less water than they used to and have begun to melt at least two weeks earlier in the spring. This earlier melting alters glacier stability, as well as downstream water supplies, with implications for wildlife, agriculture, and fire management.

In a recent study, scientists looked at 39 glaciers in and around GNP and compared aerial photos and digital maps from 1966 to 2016. Currently, only 26 glaciers are bigger than 25 acres, the minimum size used for defining a glacier. When GNP was established early in 1910, it is estimated that there were 150 glaciers larger than 25 acres. Long-term studies of glacier size have shown that the rate of melting has fluctuated in response to decade-long climate cycles and that the melting rate has risen steeply since about 1980.[47,48] Over the next 30 years, glaciologists project that most glaciers in GNP will melt to a point where they are too small to be active glaciers, and some may disappear completely. All glaciers in the park are under severe threat of completely melting by the end of the century.[4]

How are the oceans affected by climate change?

The oceans have absorbed over 90% of the excess heat energy and more than 25% of the carbon dioxide (CO_2) that is trapped in the atmosphere as a result of human-produced greenhouse gases (GHGs). Due to this increase in GHGs in the atmosphere, all ocean basins are warming and experiencing changes in their circulation and seawater chemistry, all of which alter ecosystem structure and marine biodiversity.

The world's oceans have been and will continue to be impacted by climate change. More than 50% of the world's marine ecosystems are already exposed to conditions (temperature, oxygen, salinity, and pH) that are outside the normal range of natural climate variability, and this percentage will rise as the planet warms (Ch. 9: Oceans).[1] Global warming will alter the ability of species to survive and can reorganize ecosystems, creating novel habitats and/or reducing biodiversity. Some species are responding to increased ocean temperatures by shifting their geographic ranges, generally to higher latitudes, or altering the timing of life stages (for example, spawning; Figure A5.29) (see Ch. 7: Ecosystems; Ch. 18: Northeast).[49] Other species are unable to adapt as their habitats deteriorate (for example, due to loss of sea ice) or the rate of climate-related changes occurs faster than they can move (for example, in the case of sessile organisms, such as oysters and corals).

Physical changes to the ocean system will also occur. Observations and projections suggest that in the next 100 years, the Gulf Stream (part of the larger "ocean conveyor belt") could slow down as a result of climate change, which could increase regional sea level rise and alter weather patterns along the U.S. East Coast.[13,50]

In addition to causing changes in temperature, precipitation, and circulation, increasing atmospheric levels of CO_2 have a direct effect on ocean chemistry. The oceans currently absorb about a quarter of the 10 billion tons of CO_2 emitted to the atmosphere by human activities every year. Dissolved CO_2 reacts with seawater to make it more acidic. This acidification impacts marine life such as shellfish and corals, making it more difficult for these calcifying animals to make their hard external structures (Ch. 8: Oceans; Ch. 24: Northwest).

Over the last 50 years, inland seas, estuaries, and coastal and open oceans have all experienced major oxygen losses. A warmer ocean holds less oxygen. Warming also changes the physical mixing of ocean waters (for example, upwelling and circulation) and can interact with other human-induced changes. For example, fertilizer runoff entering the Gulf of Mexico through the Mississippi River can stimulate harmful algal blooms. These blooms eventually decay, creating large "dead zones" of water with very low oxygen, where animals cannot survive. Warmer conditions slow down the rate at which this oxygen can be replaced, exacerbating the impact of the dead zone. These are just a few of the changes projected to occur, as detailed in Chapter 9: Oceans.

Projected Changes in Maximum Fish Catch Potential

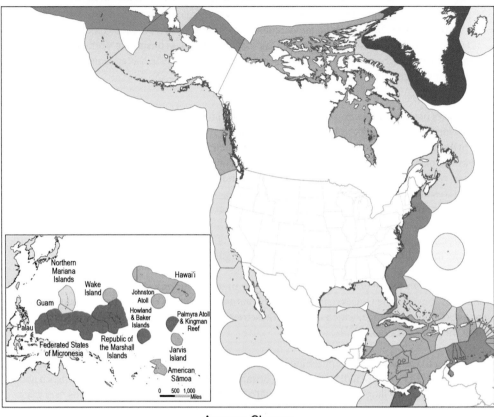

Average Change
in Maximum Catch Potential (%)

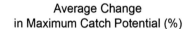

| < −30 | −30 to −20 | −20 to −10 | −10 to 0 | 0 to 10 | 10 to 20 | 20 to 30 | > 30 |

Figure A5.29: The figure shows average projected changes in fishery catches within large marine ecosystems for 2041–2060 relative to 1991–2010 under a higher scenario (RCP8.5). All U.S. large marine ecosystems, with the exception of the Alaska Arctic, are expected to see declining fishery catches. Source: adapted from Lam et al. 2016.[51]

What is ocean acidification, and how does it affect marine life?

The oceans currently absorb more than a quarter of the 10 billion tons of carbon dioxide (CO_2) released annually into the atmosphere from human activities. CO_2 reacts with seawater to form carbonic acid, so more dissolved CO_2 increases the acidity of ocean waters. When seawater reaches a certain acidity, it eats away at, or corrodes, the shells and skeletons made by shellfish, corals, and other species—or impedes the ability of organisms to grow them in the first place.

Since the beginning of the Industrial Revolution, the acidity of surface ocean waters has increased approximately 30%. The oceans will continue to absorb CO_2 produced by human activities, causing acidity to rise further (Figure A5.30). Ocean waters are not acidifying at the same rate around the globe, largely due to differences in ocean temperature. Warmer, low-latitude waters naturally hold less CO_2 and therefore tend to be less acidic. Colder, high-latitude waters naturally hold more CO_2, have increased acidity, and are closer to the threshold where shells and skeletons tend to corrode. Coastal and estuarine waters are also acidified by local phenomena, such as freshwater runoff from land, nutrient pollution, and upwelling.[1]

In the past five years, scientists have found that the shells of small planktonic snails (called pteropods) are already partially dissolved in locations where ocean acidification has made ocean waters corrosive, such as in the Pacific Northwest and near Antarctica. Pteropods are an important food source for Pacific salmon, so impacts to pteropods could cause changes up the food chain. Acidification has also affected commercial oyster hatcheries in the Pacific Northwest, where acidified waters impaired the growth and survival of oyster larvae (Ch. 24: Northwest).

Because marine species vary in their sensitivity to ocean acidification, scientists expect some species to decline and others to increase in abundance in response to this environmental change. Relative changes in species performance can ripple through the food web, reorganizing ecosystems as the balance between predators and prey shifts and habitat-forming species increase or decline. Habitat-forming species, such as corals and oysters, that grow by using minerals from the seawater to build mass are particularly vulnerable. It is difficult to predict exactly how ocean acidification will change ecosystems. Scientists and managers are now using computer models to project potential consequences to fisheries, protected species, and habitats (see Ch. 9: Oceans for more details).

Projected Change in Surface Ocean Acidity

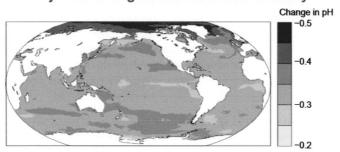

Change in pH

Figure A5.30: This figure shows projected changes in sea surface pH in 2090–2099 relative to 1990–1999 under the higher scenario (RCP8.5). As shown in the figure, every ocean is expected to increase in acidity, with increases in the Arctic Ocean projected to become the most pronounced. Source: adapted from Bopp et al. 2013 (CC BY 3.0).[52]

How do higher carbon dioxide concentrations affect plant communities and crops?

Plant communities and crops respond to higher atmospheric carbon dioxide concentrations in multiple ways. Some plant species are more responsive to changes in carbon dioxide than others, which makes projecting changes difficult at the plant community level. For approximately 95% of all plant species, an increase in carbon dioxide represents an increase in a necessary resource and could stimulate growth, assuming other factors like water and nutrients are not limiting and temperatures remain in a suitable growing range.

Along with water, nutrients, and sunlight, carbon dioxide (CO_2) is one of four resources necessary for plants to grow. At the level of a single plant, all else being equal, an increase in CO_2 will tend to accelerate growth because of accelerated photosynthesis, but a plant's ability to respond to increased CO_2 may be limited by soil nutrients. Exactly how much growth stimulation will occur varies significantly from species to species. However, the interaction between plants and their surrounding environment complicates the relationship. As CO_2 increases, some species may respond to a higher degree and become more competitive, which may lead to changes in plant community composition. For example, loblolly pine and poison ivy both grow in response to elevated CO_2; however, poison ivy responds more and becomes more competitive.[53]

The expected effects of increased CO_2 in agricultural plants are in line with these same patterns. Some crops that are not experiencing stresses from nutrients, water, or biotic stresses such as pests and disease are expected to benefit from CO_2 increases in terms of growth. However, the quality of those crops can suffer, as rising levels of atmospheric CO_2 can decrease dietary iron and other micronutrients (Ch. 14: Human Health). Plants often become less water stressed as CO_2 levels increase, because high atmospheric CO_2 allows plants to photosynthesize with lower water losses and higher water-use efficiencies. The magnitude of the effect varies greatly from crop to crop. However, for many crops in most U.S. regions, the benefits will likely be mostly or completely offset by increased stresses, such as higher temperatures, worsening air quality, and decreased ground moisture (Ch. 10: Ag & Rural). If crops and weeds are competing, then rising CO_2, in general, is more likely to stimulate the weed than the crop, with negative effects on production unless weeds are controlled.[54] Controlling weeds, however, is slightly more difficult, as rising CO_2 can reduce the efficacy of herbicides through enhanced gene transfer between crops and weedy relatives.[54]

Downstream impacts of rising CO_2 on plants can be significant. Increasing CO_2 concentrations provide an opportunity for cultivators to select plants that can exploit the higher CO_2 conditions and convert it to additional seed yield.[55] However, an area of emerging science suggests that rising CO_2 can reduce the nutritional quality (protein and micronutrients) of major crops.[56] In addition, rising CO_2 can reduce the protein concentration of pollen sources for bees.[57] Climate change also influences the amount and timing of pollen production. Increased CO_2 and temperature are correlated with earlier and greater pollen production and a longer allergy season (Ch. 13: Air Quality).

Please see Chapter 10: Ag & Rural, Chapter 6: Forests, and Ziska et al. (2016)[56] for more information on how climate change affects crops and plants.

Is climate change affecting U.S. wildfires?

It is difficult to determine how much of a role climate change has played in affecting recent wildfire activity in the United States. However, climate is generally considered to be a major driver of wildfire area burned. Over the last century, wildfire area burned in the mountainous areas of the western United States was greater during periods of low precipitation, drought, and high temperatures. Increased temperatures and drought severity with climate change will likely lead to increased fire area burned in fire-prone regions of the United States.

Climate is a major determinant of vegetation composition and productivity, which directly affect the type, amount, and structure of fuel available for fires. Climate also affects fuel moisture and the length of the season when fires are likely. Higher temperatures and lower precipitation result in lower fuel moisture, making fire spread more likely when an ignition occurs (if fuel is available). In mountainous areas, higher temperatures, lower snowpack, and earlier snowmelt lead to a longer fire season, lower fuel moisture, and higher likelihood of large fires.[58,59] Forest management practices are also a factor in determining the likelihood of ignition, as well as fire duration, extent, and intensity (Ch. 6: Forests).[23]

Long records of fire provided by tree-ring and charcoal evidence show that climate is the primary driver of fire on timescales ranging from years to millennia.[60] During the 20th century in the western United States, warm and dry conditions in spring and summer generally led to greater area burned in most places, particularly more mountainous and northerly locations (Figure A5.31).[60] The frequency of large forest fires (greater than 990 acres) has increased since the 1970s in the Northwest (1,000%) and Northern Rocky Mountains (889%), followed by forests in the Southwest (462%), Southern Rocky Mountains (274%), and Sierra Nevada (256%).[59] Dry forests in these regions account for about half of the total forest area burned since 1984. Globally, the length of the fire season (the time of year when climate and weather conditions are conducive to fire) has increased by 19% between 1979 and 2013, and it has become significantly longer over this period in most of the United States.[61]

With climate change, higher temperatures and more severe drought will likely lead to increased area burned in many ecosystems of the western and southeastern United States. By the mid-21st century, annual area burned is expected to increase 200%–300% in the contiguous western United States and 30% in the southeastern United States.[62] Over time, warmer temperatures and increased area burned can alter vegetation composition and productivity, which in turn affect fire occurrence. In arid regions, vegetation productivity may decrease sufficiently that fire will become less frequent. In other regions, climate may become less of a limiting factor for fire, and fuels may become more important in determining fire severity and extent.[63] In a warmer climate, wildfire is expected to be a catalyst for ecosystem change in all fire-prone ecosystems.

Area Burned by Large Wildfires Has Increased

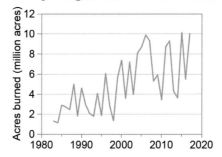

Figure A5.31: The figure shows the annual area burned by wildfires in the United States from 1983 to 2017. Warmer and drier conditions have contributed to an increase in large forest fires in the western United States and interior Alaska over the past several decades, and the ten years with the largest area burned have all occurred since 2000. Source: adapted from EPA 2016.[64]

Does climate change increase the spread of mosquitoes or ticks?

Yes. Climate change can contribute to the spread of mosquitoes and ticks. A warmer climate enhances the suitability of habitats that were formerly too cold to support mosquito and tick populations, thus allowing these vectors, and the diseases they transmit, to invade new areas.

Mosquitoes and ticks are dependent on external sources for body heat, thus they develop from egg to adult more quickly under warmer conditions, producing more generations in a shorter time. Warming also speeds up population growth of the parasites and pathogens that mosquitoes transmit (including the agents of Zika virus, dengue fever, West Nile virus, and malaria), as well as the rate at which mosquitoes bite people and other hosts. Additionally, warmer conditions facilitate the spread of mosquitoes by increasing the length of the growing season and by decreasing the likelihood of winter die-offs due to extreme cold (Ch. 14: Human Health).[65]

Blacklegged (deer) ticks are the main vector (or transmitter) of Lyme disease in the United States. These ticks require a minimum number of days above freezing to persist. As a result, some northern and high-elevation areas cannot be invaded because the warm season is too short to allow each life stage to find an animal host before it needs to retreat underground. But as higher-latitude and higher-altitude areas continue to warm, blacklegged ticks may expand their range northward and higher in elevation (Figure A5.32) (see also Ch. 14: Human Health).[66,67] Studies show that ticks emerge earlier in the spring under warmer conditions, suggesting that the main Lyme disease season will move earlier in the spring.[65] Thus, earlier onset of warm spring conditions and warm summers and falls increase the establishment and resilience of tick populations.

Lyme Disease Cases Increase Under Warmer Conditions

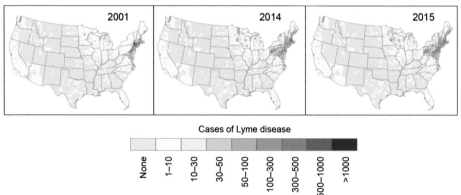

Figure A5.32: Reported cases of Lyme disease in 2001, 2014, and 2015 are shown by county for the contiguous United States. Both the distribution and total number of cases have increased from 2001 to 2014 and 2015, particularly in the Midwest and Northeast. Sources: CDC and ERT, Inc.

Acknowledgments

Technical Contributors

C. Taylor Armstrong
National Oceanic and Atmospheric Administration

Edward Blanchard-Wrigglesworth
University of Washington

James Bradbury
Georgetown Climate Center

Delavane Diaz
Electric Power Research Institute

Joshua Graff-Zivin
University of California, San Diego

Jessica Halofsky
University of Washington

Lesley Jantarasami
Oregon Department of Energy

Shannon LaDeau
Cary Institute of Ecosystem Studies

Elizabeth Marino
Oregon State University

Shaima Nasiri
U.S. Department of Energy

Matthew Neidell
Columbia University

Rachel Novak
U.S. Department of the Interior

Rick Ostfeld
Cary Institute of Ecosystem Studies

David Pierce
Scripps Institute of Oceanography

Catherine Pollack
National Oceanic and Atmospheric Administration

William V. Sweet
National Oceanic and Atmospheric Administration

Carina Wyborn
University of Montana

Laurie Yung
University of Montana–Missoula

Lewis Ziska
U.S. Department of Agriculture

References

1. Jewett, L. and A. Romanou, 2017: Ocean acidification and other ocean changes. Climate Science Special Report: Fourth National Climate Assessment, Volume I. Wuebbles, D.J., D.W. Fahey, K.A. Hibbard, D.J. Dokken, B.C. Stewart, and T.K. Maycock, Eds. U.S. Global Change Research Program, Washington, DC, USA, 364-392. http://dx.doi.org/10.7930/J0QV3JQB

2. Taylor, P.C., W. Maslowski, J. Perlwitz, and D.J. Wuebbles, 2017: Arctic changes and their effects on Alaska and the rest of the United States. Climate Science Special Report: Fourth National Climate Assessment, Volume I. Wuebbles, D.J., D.W. Fahey, K.A. Hibbard, D.J. Dokken, B.C. Stewart, and T.K. Maycock, Eds. U.S. Global Change Research Program, Washington, DC, USA, 303-332. http://dx.doi.org/10.7930/J00863GK

3. Easterling, D.R., K.E. Kunkel, J.R. Arnold, T. Knutson, A.N. LeGrande, L.R. Leung, R.S. Vose, D.E. Waliser, and M.F. Wehner, 2017: Precipitation change in the United States. Climate Science Special Report: Fourth National Climate Assessment, Volume I. Wuebbles, D.J., D.W. Fahey, K.A. Hibbard, D.J. Dokken, B.C. Stewart, and T.K. Maycock, Eds. U.S. Global Change Research Program, Washington, DC, USA, 207-230. http://dx.doi.org/10.7930/J0H993CC

4. Wuebbles, D.J., D.R. Easterling, K. Hayhoe, T. Knutson, R.E. Kopp, J.P. Kossin, K.E. Kunkel, A.N. LeGrande, C. Mears, W.V. Sweet, P.C. Taylor, R.S. Vose, and M.F. Wehner, 2017: Our globally changing climate. Climate Science Special Report: Fourth National Climate Assessment, Volume I. Wuebbles, D.J., D.W. Fahey, K.A. Hibbard, D.J. Dokken, B.C. Stewart, and T.K. Maycock, Eds. U.S. Global Change Research Program, Washington, DC, USA, 35-72. http://dx.doi.org/10.7930/J08S4N35

5. Fahey, D.W., S. Doherty, K.A. Hibbard, A. Romanou, and P.C. Taylor, 2017: Physical drivers of climate change. Climate Science Special Report: Fourth National Climate Assessment, Volume I. Wuebbles, D.J., D.W. Fahey, K.A. Hibbard, D.J. Dokken, B.C. Stewart, and T.K. Maycock, Eds. U.S. Global Change Research Program, Washington, DC, USA, 73-113. http://dx.doi.org/10.7930/J0513WCR

6. Walsh, J., D. Wuebbles, K. Hayhoe, J. Kossin, K. Kunkel, G. Stephens, P. Thorne, R. Vose, M. Wehner, J. Willis, D. Anderson, V. Kharin, T. Knutson, F. Landerer, T. Lenton, J. Kennedy, and R. Somerville, 2014: Appendix 4: Frequently asked questions. Climate Change Impacts in the United States: The Third National Climate Assessment. Melillo, J.M., Terese (T.C.) Richmond, and G.W. Yohe, Eds. U.S. Global Change Research Program, Washington, DC, 790-820. http://dx.doi.org/10.7930/J0G15XS3

7. Mann, M.E., Z. Zhang, M.K. Hughes, R.S. Bradley, S.K. Miller, S. Rutherford, and F. Ni, 2008: Proxy-based reconstructions of hemispheric and global surface temperature variations over the past two millennia. Proceedings of the National Academy of Sciences of the United States of America, 105 (36), 13252-13257. http://dx.doi.org/10.1073/pnas.0805721105

8. EPA, 2016: Climate Change Indicators in the United States, 2016. 4th edition. EPA 430-R-16-004. U.S. Environmental Protection Agency, Washington, DC, 96 pp. https://www.epa.gov/sites/production/files/2016-08/documents/climate_indicators_2016.pdf

9. USGCRP, 2017: Climate Science Special Report: Fourth National Climate Assessment, Volume I. Wuebbles, D.J., D.W. Fahey, K.A. Hibbard, D.J. Dokken, B.C. Stewart, and T.K. Maycock, Eds. U.S. Global Change Research Program, Washington, DC, 470 pp. http://dx.doi.org/10.7930/J0J964J6

10. Masson-Delmotte, V., M. Schulz, A. Abe-Ouchi, J. Beer, A. Ganopolski, J.F. González Rouco, E. Jansen, K. Lambeck, J. Luterbacher, T. Naish, T. Osborn, B. Otto-Bliesner, T. Quinn, R. Ramesh, M. Rojas, X. Shao, and A. Timmermann, 2013: Information from paleoclimate archives. Climate Change 2013: The Physical Science Basis. Contribution of Working Group I to the Fifth Assessment Report of the Intergovernmental Panel on Climate Change. Stocker, T.F., D. Qin, G.-K. Plattner, M. Tignor, S.K. Allen, J. Boschung, A. Nauels, Y. Xia, V. Bex, and P.M. Midgley, Eds. Cambridge University Press, Cambridge, United Kingdom and New York, NY, USA, 383-464. http://www.climatechange2013.org/report/full-report/

11. EPA, 2017: Climate Change Indicators: Atmospheric Concentrations of Greenhouse Gases. U.S. Environmental Protection Agency (EPA), Washington, DC. https://www.epa.gov/climate-indicators/climate-change-indicators-atmospheric-concentrations-greenhouse-gases

12. Myhre, G., D. Shindell, F.-M. Bréon, W. Collins, J. Fuglestvedt, J. Huang, D. Koch, J.-F. Lamarque, D. Lee, B. Mendoza, T. Nakajima, A. Robock, G. Stephens, T. Takemura, and H. Zhang, 2013: Anthropogenic and natural radiative forcing. Climate Change 2013: The Physical Science Basis. Contribution of Working Group I to the Fifth Assessment Report of the Intergovernmental Panel on Climate Change. Stocker, T.F., D. Qin, G.-K. Plattner, M. Tignor, S.K. Allen, J. Boschung, A. Nauels, Y. Xia, V. Bex, and P.M. Midgley, Eds. Cambridge University Press, Cambridge, United Kingdom and New York, NY, USA, 659–740. http://www.climatechange2013.org/report/full-report/

13. Perlwitz, J., T. Knutson, J.P. Kossin, and A.N. LeGrande, 2017: Large-scale circulation and climate variability. Climate Science Special Report: Fourth National Climate Assessment, Volume I. Wuebbles, D.J., D.W. Fahey, K.A. Hibbard, D.J. Dokken, B.C. Stewart, and T.K. Maycock, Eds. U.S. Global Change Research Program, Washington, DC, USA, 161-184. http://dx.doi.org/10.7930/J0RV0KVQ

14. Knutson, T., J.P. Kossin, C. Mears, J. Perlwitz, and M.F. Wehner, 2017: Detection and attribution of climate change. Climate Science Special Report: Fourth National Climate Assessment, Volume I. Wuebbles, D.J., D.W. Fahey, K.A. Hibbard, D.J. Dokken, B.C. Stewart, and T.K. Maycock, Eds. U.S. Global Change Research Program, Washington, DC, USA, 114-132. http://dx.doi.org/10.7930/J01834ND

15. Knutson, T.R., R. Zhang, and L.W. Horowitz, 2016: Prospects for a prolonged slowdown in global warming in the early 21st century. Nature Communications, 7, 13676. http://dx.doi.org/10.1038/ncomms13676

16. Peterson, T.C., W.M. Connolley, and J. Fleck, 2008: The myth of the 1970s global cooling scientific consensus. Bulletin of the American Meteorological Society, 89 (9), 1325-1338. http://dx.doi.org/10.1175/2008bams2370.1

17. DeAngelo, B., J. Edmonds, D.W. Fahey, and B.M. Sanderson, 2017: Perspectives on climate change mitigation. Climate Science Special Report: Fourth National Climate Assessment, Volume I. Wuebbles, D.J., D.W. Fahey, K.A. Hibbard, D.J. Dokken, B.C. Stewart, and T.K. Maycock, Eds. U.S. Global Change Research Program, Washington, DC, USA, 393-410. http://dx.doi.org/10.7930/J0M32SZG

18. Hayhoe, K., J. Edmonds, R.E. Kopp, A.N. LeGrande, B.M. Sanderson, M.F. Wehner, and D.J. Wuebbles, 2017: Climate models, scenarios, and projections. Climate Science Special Report: Fourth National Climate Assessment, Volume I. Wuebbles, D.J., D.W. Fahey, K.A. Hibbard, D.J. Dokken, B.C. Stewart, and T.K. Maycock, Eds. U.S. Global Change Research Program, Washington, DC, USA, 133-160. http://dx.doi.org/10.7930/J0WH2N54

19. Hawkins, E. and R. Sutton, 2009: The potential to narrow uncertainty in regional climate predictions. Bulletin of the American Meteorological Society, 90 (8), 1095-1107. http://dx.doi.org/10.1175/2009BAMS2607.1

20. Vose, R.S., D. Arndt, V.F. Banzon, D.R. Easterling, B. Gleason, B. Huang, E. Kearns, J.H. Lawrimore, M.J. Menne, T.C. Peterson, R.W. Reynolds, T.M. Smith, C.N. Williams, and D.L. Wuertz, 2012: NOAA's merged land-ocean surface temperature analysis. Bulletin of the American Meteorological Society, 93, 1677-1685. http://dx.doi.org/10.1175/BAMS-D-11-00241.1

21. Trenberth, K.E., 2015: Has there been a hiatus? Science, 349 (6249), 691-692. http://dx.doi.org/10.1126/science.aac9225

22. Kossin, J.P., T. Hall, T. Knutson, K.E. Kunkel, R.J. Trapp, D.E. Waliser, and M.F. Wehner, 2017: Extreme storms. Climate Science Special Report: Fourth National Climate Assessment, Volume I. Wuebbles, D.J., D.W. Fahey, K.A. Hibbard, D.J. Dokken, B.C. Stewart, and T.K. Maycock, Eds. U.S. Global Change Research Program, Washington, DC, USA, 257-276. http://dx.doi.org/10.7930/J07S7KXX

23. Wehner, M.F., J.R. Arnold, T. Knutson, K.E. Kunkel, and A.N. LeGrande, 2017: Droughts, floods, and wildfires. Climate Science Special Report: Fourth National Climate Assessment, Volume I. Wuebbles, D.J., D.W. Fahey, K.A. Hibbard, D.J. Dokken, B.C. Stewart, and T.K. Maycock, Eds. U.S. Global Change Research Program, Washington, DC, USA, 231-256. http://dx.doi.org/10.7930/J0CJ8BNN

24. Williams, A.P., R. Seager, J.T. Abatzoglou, B.I. Cook, J.E. Smerdon, and E.R. Cook, 2015: Contribution of anthropogenic warming to California drought during 2012-2014. Geophysical Research Letters, 42 (16), 6819-6828. http://dx.doi.org/10.1002/2015GL064924

25. Knutson, T.R., J.J. Sirutis, G.A. Vecchi, S. Garner, M. Zhao, H.-S. Kim, M. Bender, R.E. Tuleya, I.M. Held, and G. Villarini, 2013: Dynamical downscaling projections of twenty-first-century Atlantic hurricane activity: CMIP3 and CMIP5 model-based scenarios. Journal of Climate, 27 (17), 6591-6617. http://dx.doi.org/10.1175/jcli-d-12-00539.1

26. Diaz, D. and F. Moore, 2017: Quantifying the economic risks of climate change. Nature Climate Change, 7, 774-782. http://dx.doi.org/10.1038/nclimate3411

27. Rose, S.K., D.B. Diaz, and G.J. Blanford, 2017: Understanding the social cost of carbon: A model diagnostic and inter-comparison study. Climate Change Economics, 08 (02), 1750009. http://dx.doi.org/10.1142/s2010007817500099

28. National Academies of Sciences Engineering and Medicine, 2017: Valuing Climate Damages: Updating Estimation of the Social Cost of Carbon Dioxide. The National Academies Press, Washington, DC, 280 pp. http://dx.doi.org/10.17226/24651

29. Sanderson, B.M., B.C. O'Neill, and C. Tebaldi, 2016: What would it take to achieve the Paris temperature targets? Geophysical Research Letters, 43 (13), 7133-7142. http://dx.doi.org/10.1002/2016GL069563

30. Deschênes, O. and M. Greenstone, 2011: Climate change, mortality, and adaptation: Evidence from annual fluctuations in weather in the US. American Economic Journal: Applied Economics, 3 (4), 152-185. http://dx.doi.org/10.1257/app.3.4.152

31. Graff Zivin, J. and M. Neidell, 2014: Temperature and the allocation of time: Implications for climate change. Journal of Labor Economics, 32 (1), 1-26. http://dx.doi.org/10.1086/671766

32. Schlenker, W. and M.J. Roberts, 2009: Nonlinear temperature effects indicate severe damages to U.S. crop yields under climate change. Proceedings of the National Academy of Sciences of the United States of America, 106 (37), 15594-15598. http://dx.doi.org/10.1073/pnas.0906865106

33. Adhvaryu, A., N. Kala, and A. Nyshadham, 2014: The Light and the Heat: Productivity Co-benefits of Energy-Saving Technology. NBER Working Paper No. 24314. National Bureau of Economic Research, Cambridge, MA, 63 pp. http://dx.doi.org/10.3386/w24314

34. Hsiang, S.M. and A.S. Jina, 2014: The Causal Effect of Environmental Catastrophe on Long-Run Economic Growth: Evidence from 6,700 Cyclones. NBER Working Paper No. 20352. National Bureau of Economic Research, Cambridge, MA, 68 pp. http://dx.doi.org/10.3386/w20352

35. Franco, G. and A.H. Sanstad, 2008: Climate change and electricity demand in California. Climatic Change, 87 (Suppl. 1), 139-151. http://dx.doi.org/10.1007/s10584-007-9364-y

36. NAS, 2015: Climate Intervention: Carbon Dioxide Removal and Reliable Sequestration. The National Academies Press, Washington, DC, 154 pp. http://dx.doi.org/10.17226/18805

37. NAS, 2015: Climate Intervention: Reflecting Sunlight to Cool Earth. The National Academies Press, Washington, DC, 260 pp. http://dx.doi.org/10.17226/18988

38. Shepherd, J., K. Caldeira, P. Cox, J. Haigh, D. Keith, B. Launder, G. Mace, G. MacKerron, J. Pyle, S. Rayner, C. Redgwell, and A. Watson, 2009: Geoengineering the Climate: Science, Governance and Uncertainty. Report 10/09. The Royal Society, London, UK, 82 pp. https://royalsociety.org/~/media/Royal_Society_Content/policy/publications/2009/8693.pdf

39. Sweet, W.V., R. Horton, R.E. Kopp, A.N. LeGrande, and A. Romanou, 2017: Sea level rise. Climate Science Special Report: Fourth National Climate Assessment, Volume I. Wuebbles, D.J., D.W. Fahey, K.A. Hibbard, D.J. Dokken, B.C. Stewart, and T.K. Maycock, Eds. U.S. Global Change Research Program, Washington, DC, USA, 333-363. http://dx.doi.org/10.7930/J0VM49F2

40. Sweet, W.V., R.E. Kopp, C.P. Weaver, J. Obeysekera, R.M. Horton, E.R. Thieler, and C. Zervas, 2017: Global and Regional Sea Level Rise Scenarios for the United States. NOAA Tech. Rep. NOS CO-OPS 083. National Oceanic and Atmospheric Administration, National Ocean Service, Silver Spring, MD, 75 pp. https://tidesandcurrents.noaa.gov/publications/techrpt83_Global_and_Regional_SLR_Scenarios_for_the_US_final.pdf

41. NASA, 2016: Weekly Animation of Arctic Sea Ice Age with Graph of Ice Age by Area: 1984-2016. NASA Scientific Visualization Studio, accessed 12 February. https://svs.gsfc.nasa.gov/4510

42. Stroeve, J. and D. Notz, 2015: Insights on past and future sea-ice evolution from combining observations and models. Global and Planetary Change, 135, 119-132. http://dx.doi.org/10.1016/j.gloplacha.2015.10.011

43. Wouters, B., J.L. Bamber, M.R. van den Broeke, J.T.M. Lenaerts, and I. Sasgen, 2013: Limits in detecting acceleration of ice sheet mass loss due to climate variability. Nature Geoscience, 6 (8), 613-616. http://dx.doi.org/10.1038/ngeo1874

44. Pederson, G.T., L.J. Graumlich, D.B. Fagre, T. Kipfer, and C.C. Muhlfeld, 2010: A century of climate and ecosystem change in Western Montana: What do temperature trends portend? Climatic Change, 98 (1-2), 133-154. http://dx.doi.org/10.1007/s10584-009-9642-y

45. Pederson, G.T., S.T. Gray, C.A. Woodhouse, J.L. Betancourt, D.B. Fagre, J.S. Littell, E. Watson, B.H. Luckman, and L.J. Graumlich, 2011: The unusual nature of recent snowpack declines in the North American cordillera. Science, 333 (6040), 332-335. http://dx.doi.org/10.1126/science.1201570

46. Pederson, G.T., J.L. Betancourt, and G.J. McCabe, 2013: Regional patterns and proximal causes of the recent snowpack decline in the Rocky Mountains, U.S. Geophysical Research Letters, 40 (9), 1811-1816. http://dx.doi.org/10.1002/grl.50424

47. Pederson, G.T., D.B. Fagre, S.T. Gray, and L.J. Graumlich, 2004: Decadal-scale climate drivers for glacial dynamics in Glacier National Park, Montana, USA. Geophysical Research Letters, 31 (12), L12203. http://dx.doi.org/10.1029/2004GL019770

48. Pederson, G.T., S.T. Gray, T. Ault, W. Marsh, D.B. Fagre, A.G. Bunn, C.A. Woodhouse, and L.J. Graumlich, 2011: Climatic controls on the snowmelt hydrology of the northern Rocky Mountains. Journal of Climate, 24 (6), 1666-1687. http://dx.doi.org/10.1175/2010jcli3729.1

49. Burrows, M.T., D.S. Schoeman, L.B. Buckley, P. Moore, E.S. Poloczanska, K.M. Brander, C. Brown, J.F. Bruno, C.M. Duarte, B.S. Halpern, J. Holding, C.V. Kappel, W. Kiessling, M.I. O'Connor, J.M. Pandolfi, C. Parmesan, F.B. Schwing, W.J. Sydeman, and A.J. Richardson, 2011: The pace of shifting climate in marine and terrestrial ecosystems. Science, 334, 652-655. http://dx.doi.org/10.1126/science.1210288

50. Yang, H., G. Lohmann, W. Wei, M. Dima, M. Ionita, and J. Liu, 2016: Intensification and poleward shift of subtropical western boundary currents in a warming climate. Journal of Geophysical Research Oceans, 121 (7), 4928-4945. http://dx.doi.org/10.1002/2015JC011513

51. Lam, V.W.Y., W.W.L. Cheung, G. Reygondeau, and U.R. Sumaila, 2016: Projected change in global fisheries revenues under climate change. Scientific Reports, 6, Art. 32607. http://dx.doi.org/10.1038/srep32607

52. Bopp, L., L. Resplandy, J.C. Orr, S.C. Doney, J.P. Dunne, M. Gehlen, P. Halloran, C. Heinze, T. Ilyina, R. Séférian, J. Tjiputra, and M. Vichi, 2013: Multiple stressors of ocean ecosystems in the 21st century: Projections with CMIP5 models. Biogeosciences, 10 (10), 6225-6245. http://dx.doi.org/10.5194/bg-10-6225-2013

53. Mohan, J.E., L.H. Ziska, W.H. Schlesinger, R.B. Thomas, R.C. Sicher, K. George, and J.S. Clark, 2006: Biomass and toxicity responses of poison ivy (Toxicodendron radicans) to elevated atmospheric CO_2. Proceedings of the National Academy of Sciences of the United States of America, 103 (24), 9086-9089. http://dx.doi.org/10.1073/pnas.0602392103

54. Ziska, L.H., D.R. Gealy, M.B. Tomecek, A.K. Jackson, and H.L. Black, 2012: Recent and projected increases in atmospheric CO_2 concentration can enhance gene flow between wild and genetically altered rice (Oryza sativa). PLOS ONE, 7 (5), e37522. http://dx.doi.org/10.1371/journal.pone.0037522

55. Taub, D.R., 2010: Effects of rising atmospheric concentrations of carbon dioxide on plants. Nature Education Knowledge, 3 (10), 21. https://www.nature.com/scitable/knowledge/library/effects-of-rising-atmospheric-concentrations-of-carbon-13254108

56. Ziska, L., A. Crimmins, A. Auclair, S. DeGrasse, J.F. Garofalo, A.S. Khan, I. Loladze, A.A. Pérez de León, A. Showler, J. Thurston, and I. Walls, 2016: Ch. 7: Food safety, nutrition, and distribution. The Impacts of Climate Change on Human Health in the United States: A Scientific Assessment. U.S. Global Change Research Program, Washington, DC, 189-216. http://dx.doi.org/10.7930/J0ZP4417

57. Ziska, L.H., J.S. Pettis, J. Edwards, J.E. Hancock, M.B. Tomecek, A. Clark, J.S. Dukes, I. Loladze, and H.W. Polley, 2016: Rising atmospheric CO_2 is reducing the protein concentration of a floral pollen source essential for North American bees. Proceedings of the Royal Society B: Biological Sciences, 283 (1828). http://dx.doi.org/10.1098/rspb.2016.0414

58. Barbero, R., J.T. Abatzoglou, N.K. Larkin, C.A. Kolden, and B. Stocks, 2015: Climate change presents increased potential for very large fires in the contiguous United States. International Journal of Wildland Fire. http://dx.doi.org/10.1071/WF15083

59. Westerling, A.L., 2016: Increasing western US forest wildfire activity: Sensitivity to changes in the timing of spring. Philosophical Transactions of the Royal Society B: Biological Sciences, 371, 20150178. http://dx.doi.org/10.1098/rstb.2015.0178

60. Littell, J.S., D. McKenzie, D.L. Peterson, and A.L. Westerling, 2009: Climate and wildfire area burned in western U.S. ecoprovinces, 1916-2003. Ecological Applications, 19 (4), 1003-1021. http://dx.doi.org/10.1890/07-1183.1

61. Jolly, W.M., M.A. Cochrane, P.H. Freeborn, Z.A. Holden, T.J. Brown, G.J. Williamson, and D.M.J.S. Bowman, 2015: Climate-induced variations in global wildfire danger from 1979 to 2013. Nature Communications, 6, 7537. http://dx.doi.org/10.1038/ncomms8537

62. Prestemon, J.P., U. Shankar, A. Xiu, K. Talgo, D. Yang, E. Dixon, D. McKenzie, and K.L. Abt, 2016: Projecting wildfire area burned in the south-eastern United States, 2011–60. International Journal of Wildland Fire, 25 (7), 715-729. http://dx.doi.org/10.1071/WF15124

63. McKenzie, D. and J.S. Littell, 2017: Climate change and the eco-hydrology of fire: Will area burned increase in a warming western USA? Ecological Applications, 27 (1), 26-36. http://dx.doi.org/10.1002/eap.1420

64. EPA, 2016: Climate Change Indicators: Wildfires. U.S. Environmental Protection Agency (EPA), Washington, DC. https://www.epa.gov/climate-indicators/climate-change-indicators-wildfires

65. Beard, C.B., R.J. Eisen, C.M. Barker, J.F. Garofalo, M. Hahn, M. Hayden, A.J. Monaghan, N.H. Ogden, and P.J. Schramm, 2016: Ch. 5: Vector-borne diseases. The Impacts of Climate Change on Human Health in the United States: A Scientific Assessment. U.S. Global Change Research Program, Washington, DC, 129–156. http://dx.doi.org/10.7930/J0765C7V

66. Ogden, N.H., M. Radojević, X. Wu, V.R. Duvvuri, P.A. Leighton, and J. Wu, 2014: Estimated effects of projected climate change on the basic reproductive number of the Lyme disease vector Ixodes scapularis. Environmental Health Perspectives, 122, 631-638. http://dx.doi.org/10.1289/ehp.1307799

67. Ostfeld, R.S. and J.L. Brunner, 2015: Climate change and Ixodes tick-borne diseases of humans. Philosophical Transactions of the Royal Society B: Biological Sciences, 370 (1665), 20140051. http://dx.doi.org/10.1098/rstb.2014.0051